THE SENTIENT ROBOT

THE LAST TWO HURDLES IN THE RACE TO BUILD ARTIFICIAL SUPERINTELLIGENCE

Rupert Robson

imprint-academic.com

Published in the UK by
Imprint Academic, PO Box 200, Exeter EX5 5YX, UK

Distributed in the USA by
Ingram Book Company,
One Ingram Blvd., La Vergne, TN 37086, USA

ISBN 9781788360791 Paperback

A CIP catalogue record for this book is available from the
British Library and US Library of Congress

To my wife, Georgina,
and my children,
Imogen, Alexander and Christian

Contents

Part 4.
The Mirrored Homunculus Theory

Part 5.
Building the Sentient Robot

Glossary

Term	Meaning
Abduction	Abduction is about arriving at a hypothesis on the basis of just one or a few examples
Algorithm	A set of instructions. These might be hand-written or in the form of a computer program
Artificial general intelligence or AGI	Artificial intelligence with general rather than narrow intelligence capabilities. Sometimes defined as artificial human-level intelligence or AHLI
Artificial human-level intelligence or AHLI	Artificial intelligence at approximately human capability levels
Artificial intelligence or AI	Machine-based capability to compute, reason, estimate, infer. Such capabilities take the form of algorithms
Artificial neural network or ANN	A form of machine learning algorithm that emulates aspects of the human brain in its design
Artificial super-intelligence or ASI	Artificial intelligence well in excess of human capability levels
Cognition	Learning, knowing and reasoning, but excluding emotions
Computational theory of mind or CTM	The theory that the brain, especially as it relates to cognition, operates in an analogous fashion to a computer, i.e., that the brain is an information processing system
Consciousness	A term often expressed in other ways such as awareness, self-awareness and subjective experience. A computer or a thermostat, for example, processes data or inputs. A brain similarly processes inputs but this processing is also accompanied by an inner feel or feeling (not to be confused with emotion). I can ask, what does it feel like to be the President of the USA? I cannot ask, what does it feel like to be a refrigerator? A refrigerator just is. This inner feel is consciousness
Deduction	Deduction is logical inference
Deep learning	Machine learning using several or many layers of processing in an ANN
Deep neural/ learning network	ANN designed using several or many layers of processing
Emotion(s)	Complex, unconscious, automated neuronal networks that motivate animals to behave in certain ways. Emotions are typically divided between the primary and the social emotions but there is no hard and fast agreement on the categorisation of the emotions
Expert, or knowledge-based, systems	These algorithms operate in areas where an expert has been able to work out the answers to all the possible questions in advance. Alternatively, if working out all the answers would take too long, then an expert works out the rules that can come up with all the

Term	Meaning
	answers
Feelings of emotion(s)	The consciousness, or conscious awareness or experience, of an underlying emotion
General intelligence	The ability to perform deduction, induction and abduction. General intelligence includes common sense
Goal-directed VBDM	Goal-directed value-based decision-making tends to apply in novel situations or involve autonomously set goals
Hard problem	David Chalmers' term for the question, why does all the activity or neuronal processing in our brains need to feel like anything at all?
Homunculus	A very small human or humanoid creature metaphorically sitting inside the brain and directing behaviour
HPAI	Human-purposed AI, that is, AI explicitly designed so as to act in the best interests of humans
Induction	Induction is about reaching an inference that by most standards is reasonable
Instantiate	To put in or put into, to house, to locate, to install, to embed
Machine consciousness	Consciousness as instantiated in a machine such as an AI
Mentalese	The neuronal language or code that the brain employs to perform cognition
Machine learning	AI with a learning capability such that, given an initial design, the algorithm can build itself through training
Model-based learning	Model-based learning relies on mental representations, or models, of the world and possible outcomes or futures
Model-free learning	Model-free learning is narrow in scope. It relies purely on possessing an updateable capacity. This capacity can inform the brain's decision-making mechanisms as to the reward for such and such an action. The brain does not need to possess a model of the surrounding environment, a model of the world, for such a capacity to function
Model of the world	The idea that we each carry a deep, rich model of the world in our brains. This model comprises our mental representations or symbols of the objects, ideas and events we have perceived in the world. The model also consists of relationships that we have logged between these mental representations
Neocortex	The outermost layer of the brain. It includes the prefrontal cortex, which houses much of the brain's reasoning abilities
Nervous system	The body's control system, consisting in turn of the central nervous and peripheral nervous systems
Neuron, or nerve cell	The primary processing cell of the nervous system
Neuronal network	A number of neurons connected to each other and capable of processing information from sensory inputs *en route* to producing behavioural outputs
Neuroplasticity	The process whereby the brain experiences continual change throughout a person's life. It underpins learning
Neurotransmitter(s)	Chemical compounds that cross from one neuron to the other at the synapse, thereby enabling synaptic connectivity
Panpsychism	The belief that pretty much everything, down as far as sub-atomic particles such as electrons, possesses consciousness
Prediction	A term used both in neuroscience and in computer science to refer to an educated guess
Primary, or basic,	The primary emotions number roughly between five and 10 in total,

Term	Meaning
emotions	although there is as yet no hard and fast agreement on this. They probably include curiosity or seeking, fear, anger, lust, care, sadness and play or joy. They may also include disgust and pride. They are evolutionarily older than the social emotions
Probabilistic language of thought or PLoT	The theory that the brain, especially as it relates to cognition, employs both statistical and symbolic approaches to reasoning
Reinforcement learning	This describes the process whereby an algorithm corrects itself during training as a result of being programmed to maximise a certain goal such as winning points. The results that an algorithm achieves during the learning process are awarded points. It self-calibrates in order to select strategies that maximise points and eliminates strategies that do the reverse
Robot	Whilst in the broadest sense robots refer to any automated mobile machinery, in this book I use robots to refer to embodied AI
Social, or secondary, emotions	The social emotions are evolutionarily younger than the primary emotions. They probably evolved to help bind together complex mammalian species in particular such as the great apes and dolphins and whales
Statistical approach to reasoning	The theory that reasoning is a matter of probabilities based on the observation of a number of examples or data. In a human brain, the statistical approach to reasoning is reflected in System 1 or Type 1 processing. In an AI, the statistical approach to reasoning is reflected in deep learning networks
Stimulus-response VBDM	Stimulus-response value-based decision-making tends to be simpler and habitual and the decision-making process automatic, even if potentially sophisticated
Stochastic	"Having a random probability distribution or pattern that can be analysed statistically but not predicted precisely" (taken from Oxford English Dictionary; see References, chapter 9, note 18)
Supervised learning	The training of an algorithm by supervisory correction (i.e., with human input). When the algorithm errs during the learning process, it is given the correct answer and it self-calibrates accordingly. Most of today's machine learning algorithms are built using supervised learning
Symbol or symbols	Tokens standing in for, or representing, objects or ideas in the world. Mental representations are symbols embodied by neurons or networks of neurons in the brain that reflect, or represent, objects or ideas in the world that we have perceived
Symbolic approach to reasoning	The theory that reasoning is about the manipulation of mental representations by virtue of the relationships between them. In a human brain, the symbolic approach to reasoning is reflected in System 2 or Type 2 processing. In an AI, the symbolic approach to reasoning is reflected in expert systems but has as yet not been successfully emulated in deep learning networks
Synapses, or synaptic connections	The connections between neurons
System 1 or Type 1 processing	System 1 processing, or thinking, is underpinned by the forming of associations. It is probabilistic, or statistical as it is generally termed, in nature. It is fast, automatic and involves relatively little effort
System 2 or Type 2 processing	System 2 processing, or thinking, is logical rather than probabilistic. It is serial in nature. It involves effort and, as such, can easily be dominated by lazy System 1 processing

Term	Meaning
Training	The process whereby the initial design version of an algorithm self-calibrates by repeated data runs to achieve a specific goal such as image recognition or playing games. The algorithm is calibrated through learning. There are three main types of training: supervised, unsupervised and reinforcement
Two Hurdles	The Two Hurdles are the two building blocks for AGI that currently remain beyond our conceptual reach. The first of the Two Hurdles is about consciousness. The challenge set by consciousness is to accurately define it and then instantiate or embed it in AI. The second of the Two Hurdles has to do with emulating humans' flexible approach to thinking in AI so as to yield understanding and judgement
Type 1 or System 1 processing	Type 1 processing, or thinking, is underpinned by the forming of associations. It is probabilistic, or statistical as it is generally termed, in nature. It is fast, automatic and involves relatively little effort
Type 2 or System 2 processing	Type 2 processing, or thinking, is logical rather than probabilistic. It is serial in nature. It involves effort and, as such, can easily be dominated by lazy Type 1 processing
Unsupervised learning	This describes the process whereby an algorithm corrects itself during training. Unsupervised learning is a relatively sparsely used training format in machine learning
Value-based decision-making or VBDM	A type of decision-making that entails a comparison of competing options based upon a mental evaluation of each of them. VBDM is of two types: stimulus-response and goal-directed
Winner-takes-all mechanism	A neuronal mechanism by which alternative courses of action, having been created by a combination of emotion and cognition, are compared with each other such that one particular course of action iteratively succeeds in dominating the others

Introduction

Designing the sentient robot

We live in a world where change, and ever-quickening change at that, has become the norm. The climate is becoming warmer and more volatile. Politics, at least in the West, is in ferment. As I write this, in the middle of Covid-19, our daily routines are constantly up for grabs as governments ban this and reintroduce that. With another splurge of monetary easing after the QE that followed the global financial crisis of 2008, the discipline of economics is having to be rewritten. So, how the world might develop over the next 50 years or so is necessarily of acute interest. Ultimately, how it develops will be a function of how humans view themselves and their position in it. The German language possesses a useful word for this idea, *Weltanschauung*, which literally means world outlook.

Perhaps the last great shift in *Weltanschauung* was the decline in religious belief and the rise of secularism. The philosopher Charles Taylor compellingly traces this story in the West over the last 500 years in his book, *A Secular Age*.[1] The question he asks is "why was it virtually impossible not to believe in God in, say, 1500 in our Western society, while in 2000 many of us find this not only easy, but even inescapable?"[2]

We are still in the secular age, but I believe that we are at the tail end of it. The turning point, and the step into a new age, which I call the *age of equivalence*, will arise from the continuing development of artificial intelligence (AI). This is not to say that today's AI is capable of making that step. As we shall come on to see, today's AI has made great strides. Indeed, we have come to rely on it in a large number of areas of our everyday lives. It is, however, still relatively narrow in scope. Human intelligence, whilst slower in certain domains, is both broader and deeper than AI. The turning point will come when AI has advanced so far that it matches human intelligence in every way. This is often termed artificial general intelligence (AGI). Once AI has matched human intelligence, it could be but the blink of an eye before it goes on to surpass humans by a huge distance. That scary prospect is termed artificial superintelligence

(ASI). The good news is that AGI and ASI are decades away, so we still have time to prepare for their arrival.

You might well ask, "if it's really the case that AI could develop to that point, why would we permit it?" It will happen anyway despite concerns as to its safety. Perhaps the most compelling reason is that AI and the products and services enabled by AI have so far been hugely enjoyed by many, many people and useful to them too. Why would companies not continue to develop AI given the insatiable demand for it? Mankind has a consistent record of pushing the boundaries of discovery and invention, despite the risks. We have travelled to the relentlessly hostile environs of space and the depths of the oceans. We have manufactured nuclear weapons in quantities that can end life as we know it many times over. We are already well down the road towards AGI. Whilst it is not necessarily the answer to every problem in the field, machine learning has taken huge steps in just the last decade. There is no let-up in its progress.

So, let us shape AI and, thence, AGI and ASI in a way that suits us, that benefits us and that minimises the risks to us.

I suggest that in the best case we would like AI to help us, to improve us and to protect us. Towards the end of the book, I shall introduce you to Servilius, a domestic robot, who will be designed to revolutionise your home life. Servilius is just one example of how AGI could help us. There are countless others. I shall also introduce you to Felicity, an ASI oracle. She will be able to compute the answers to questions such as: what is the cure for this or that type of cancer? She will also be able to help address questions without perfect answers such as how we might limit or control climate change. On balance, I have an optimistic view of human nature—look how far we have come in the last 10,000 years. Imagine how ASI, if deployed in the right way, could add to that. Felicity will be able to improve us and our way of life at a massively accelerated rate.

And the worst case? I noted that we need AI to protect us. The West, and particularly the USA, is entering a state of active geopolitical rivalry with China. China and the USA lead the world in AI development. In China, it appears that the state is harnessing AI's capabilities to place the Chinese Communist Party's grip on power beyond any conceivable threat. It would be unthinkable for the USA, and maybe the rest of the West too, to allow China to dominate in the fields of AGI and ASI. If it does, then it may well seek to control not just its own people, but the entire world. As none other than Vladimir Putin, President of Russia, said: "Whoever becomes the leader in [artificial intelligence] will become the ruler of the world."[3]

The other manifestation of the worst case is that ASI itself threatens us. Many of us will remember Hal locking Dr David Bowman out of the spacecraft in Stanley Kubrick's film, *2001: A Space Odyssey*.[4] Interestingly, our emotional reaction was to turn against Hal as if Hal were sentient, like us. Hal had revealed his true colours: he had betrayed Dr Bowman; he was the enemy. Yet, the more likely state of Hal's computational innards was indifference in the sense of simply executing a program requirement. Hal's algorithms did not allow for Dr Bowman's re-entry.

The point is that ASI will quite likely be an agent in its own right, acting in and on the world. It may possess consciousness and it may not. Either way, if it is not wholly aligned with our interests, then it may, even if inadvertently, sweep us out of existence. Felicity will be best placed to improve us if she is fully invested in us. It is hard to grasp the sheer power of an ASI, even if it is just a concept at this stage. We absolutely need it to notice us, to be on our side and to nurture us. As Susan Schneider, another philosopher, notes: "The value that an AI places on us may hinge on whether it believes it feels like something to be us. This insight may require nothing less than machine consciousness."[5]

AGI and ASI will have the ability to compute, to calculate, to reckon, to reason, to estimate, to solve and so on. What is missing from this list? Simply put, to be human. If ASI cannot see the world wholly and exclusively through our eyes, it may ultimately not help, improve and protect us. So, we need to equip AGI and ASI with emotions, human emotions. ASI needs to know how we set goals, decide between competing priorities and generally exercise our autonomy. Even that is not enough. ASI must possess consciousness, must be aware of itself, must feel. For it will then feel like we do, feel our emotions, empathise with us and be deeply invested in us.

Until a few years ago, it was largely the world of fiction and films that contemplated the idea of AGI and ASI. Films in particular have brought ASI to the attention of the public at large. In addition to *2001: A Space Odyssey*, they include well-known titles such as *Blade Runner*,[6] *Ex Machina*[7] and *I, Robot*.[8] The world of computer science is now taking the idea of AGI and ASI more seriously, albeit without the hyperbole and drama of such movies. DeepMind, owned by Alphabet Inc., the owner of Google, and perhaps the leading AI research firm in the world, states that their "long term aim is to solve intelligence, developing more general and capable problem-solving systems, known as artificial general intelligence (AGI)".[9] Nick Bostrom, one of the world's leading thinkers on the impact of future technology on humans, has written a best-selling book on ASI called *Superintelligence: Paths, Dangers, Strategies*.[10]

The emergence of a new race of artificially superintelligent entities will mean that humans will need to make a big adjustment in their *Weltanschauung*. Prior to the secular age, we deferred to a greater power than ourselves, an entity with a higher value, God, whichever one you happened to follow. In the last 500 years, we have become used to being top dog. Giving that up to share our dominant position on Earth with an equivalent entity, ASI, will be a wrench. But, with the potential advent of ASI, that is what we will have to come to terms with doing in order to cope with the age of equivalence.

ASI is likely to be the most powerful ever invention by mankind with the most far-reaching consequences. It will happen, maybe not in my lifetime, but certainly in my children's lifetimes. I will argue that we are now just two major conceptual hurdles away from developing ASI (the Two Hurdles). These are far from trivial obstacles, but there are only two of them.

The first of the Two Hurdles is about consciousness. Clearing this hurdle will give us the sentient robot. On the face of it, there seems to be no good reason to instantiate or embed consciousness in AI. But the potential power of ASI is such that we need to think hard about how to keep it under our direct or indirect control. An important part of that will be about designing it so that it sees the world through our eyes. This is why it is essential that ASI possesses consciousness. It will not use that consciousness to the same extent or in the same way that we do. The circumscribed nature of ASI's consciousness, however, will enable it to empathise fully with us. ASI will thereby be in a position to help us rather than hurting us, even if inadvertently.

If we design ASI and the framework in which it operates correctly, then it will help, improve and protect us. The consequences of designing ASI incorrectly would be simply dreadful.

The second of the Two Hurdles is about the developmental steps needed in AI design so as to achieve human-level flexibility in thought. For example, the algorithm that recognises the image of your face in your passport cannot play chess, recommend books and films that you may like or reason logically. Humans can do all four because they possess general intelligence. At a deeper level, today's AI is unable to exhibit common sense or understanding. Today's AI is essentially statistical in nature. It lacks a model of the world, which we take for granted. This is what allows us to categorise things, to see relationships between people, objects and events and to learn rapidly from just one or two examples. Today's AI instead typically relies on large quantities of data from which it can form probabilistic associations. We can do that too, albeit not at the same speed, but we have other techniques for thinking about things too.

My approach

The book opens with the topic of artificial intelligence as we all know it and use it today, prior to homing in on the goal of human-purposed artificial superintelligence. Next up is some pertinent but accessible background in neuroscience and philosophy. Only then do we take a look at the critically important topic of consciousness. We then come down to Earth again, but an Earth decades into the future, when artificial superintelligence will be firmly part of our daily existence.

At the heart of the book lies the challenge of instantiating consciousness in a robot. This will be how we can build ASI that will see the world through our eyes. In order to meet that challenge, we will need to work out why and how consciousness came to exist in humans. I propose the mirrored homunculus theory of consciousness. The mirrored homunculus theory of consciousness is rooted firmly in an evolutionary framework. I contend that consciousness evolved like any other human feature and came to serve the purpose of promoting individual, prosocial decision-making and flexible thinking. Otherwise, it would have died out.

I suggest that, in a child's earliest, formative years, its brain mirrors the self that it perceives in its parents or caregivers. This self is the mirrored homunculus. The mirroring is facilitated by a relatively newly discovered type of neuron, the mirror neuron. The early brain perceives an agent or homunculus in its parents or caregivers by way of its mirror neurons. It thereby mirrors that, in neuronal network form, in itself—the mirrored homunculus. A homunculus is the philosophical term for a very small human or humanoid creature metaphorically sitting inside the brain and directing behaviour.

The rest of the brain interacts extensively with the mirrored homunculus as a result of the sheer reach of the mirrored homunculus neuronal network across the brain. The rest of the brain mistakenly perceives the mirrored homunculus network as an agent, which is of course an illusion. This illusion is consciousness. The brain is used to interpreting certain things as illusions. The Necker cube is an example of an optical illusion whereby the brain flicks back and forth between seeing the cube in two alternative orientations. The perception of the mirrored homunculus is also in the nature of an illusion. That illusion, consciousness, has been fundamental to humankind's success. It has elevated our decision-making to promote sociability and cooperation. It has also enabled us to think in ways that very few other animals can manage, that is, flexibly, counterfactually, causally and with common sense.

Part 5 of the book, "Building the Sentient Robot", pulls all the strands together. I introduce the reader to Servilius and to Felicity, the ASI oracle. Felicity possesses the full suite of human characteristics, including emotions, empathy and consciousness. Additionally, her cognitive powers are massively greater and quicker than those of a human. She has autonomous learning capabilities. Crucially, she possesses a suite of human values. These have been self-instantiated by virtue of, among other things, the possession of human emotions. With that range of capabilities, she is in an extraordinarily strong position to consider and formulate appropriate answers to humanity's intractable problems. She is a sentient robot who sees the world through our eyes.

Acknowledgements

I am hugely grateful to many people who have helped and encouraged me along the way. I first came up with the idea of the mirrored homunculus 10 years ago whilst walking along a beach with my wife, Georgina, in Kerala in India. I started on the first draft of my book three years later. Georgina, and my brother, Chris, were both kind enough to read that early draft and comment on it.

I give thanks to those who commented on parts or all of a subsequent draft, which appeared four years later. They included Georgina, Imogen White, Alexander Robson, Christian Robson, Alasdair McWhirter, Robert Hobhouse, Rod Banner, Judith Hussey, Iain Robertson, Justin Manson, James Stallard, Roger Emery, Imogen Pelham, Madelon Treneman, Sophie Robertson, Zara Harmon and Richard Perrin. Further helpful comments and contributions to the project were made by Cecilia Tilli, Janet Pierrehumbert, Anil Gomes and Anna Hoerder-Suabidessen. Thanks too to Alexander Robson for producing the illustrations throughout the book and to Steve Kelly for producing the front cover.

I am also deeply grateful to Imprint Academic, my publisher, and Graham Horswell, Imprint's Managing Editor. Thanks also to Jeff Scott, my publicist.

Part 1.

Human-Purposed Artificial Intelligence

Chapter 1

The Promise of AI

I woke up this morning, dressed and came downstairs for breakfast. My smart kettle was boiling the water for my green tea. Meanwhile, Alexa updated me on the day's weather and what was in my diary. It turned out that I had a free slot in the early evening. I would use that to take a virtual reality trip to Australia to establish whether I should take the family there for Christmas. I had run out of butter and my fridge added that to my shopping list. I read through the newspapers online, my iPad having automatically downloaded them. My favourite articles by reference to journalist and topic were automatically prioritised for me.

That is what is available right now, and all before leaving the house in the morning to go to work. In fact, the algorithms underpinning many such facilities are widely available too, as well as online tutorials showing you how to master them. Compare today's environment to what you could do in 2010. The difference is huge. Compare today's environment to what 2030 might look like given the pace of development in AI. The difference will again be huge.

AI's so-called algorithms, or sets of instructions, work out answers to problems. An algorithm's ability to come up with good answers depends upon its design, the data that is fed into it and its application.[1] As we shall see, today's AI is able to achieve much but it remains narrow in scope. The challenge for the continuing development of AI is not to do with its artificial nature. Rather, it has to do with the sheer scope of the word, intelligence. Human intelligence ranges across a wide spectrum of skills and attributes. It encompasses not just recognising faces, for example, but also goal-setting, planning and creativity. It encompasses the emotions that play such an important role in setting goals and in decision-making. It could even be said to encompass consciousness. The further development of artificial intelligence will involve closing the gap between where AI stands today and AI that at a minimum matches every aspect of a human's intelligence. At that point, we will have developed a tool that could, indeed should, be of profound utility and benefit to humankind.

For the time being, however, let us start by taking a closer look at where AI stands today.

AI's strengths

LawGeex is a technology firm, which specialises in solving problems in the legal world. It has developed an automated AI-powered contract review platform. In early 2018, it reported on a competition between its platform and a group of 20 US lawyers. The competition was supervised by law professors from Stanford and Duke universities.[2] The challenge was to review five so-called non-disclosure agreements (NDAs). The five NDAs consisted in aggregate of 153 paragraphs and 3,213 clauses. The average time taken by the lawyers in the competition to review the five NDAs was 92 minutes. LawGeex's platform reviewed them in 26 seconds. As for accuracy of reviews, the platform reached 94% whilst the average of the lawyers was 85%; only the best of the lawyers scored 94%.

At its best, AI can be both quick and faultless in producing answers to questions or in generating new knowledge. Once you know that an algorithm is working well, you can rely on the answers it generates. AI is tireless, never complains, never strikes, works 24/7 and, once installed, has a small marginal cost of operation.

AI first made real progress in the field of so-called expert, or knowledge-based, systems. These systems operate in areas where an expert has been able to work out the answers to all the possible questions in advance. Alternatively, if working out all the answers would take too long, then an expert works out the rules that can come up with all the answers. Those answers and/or rules together with an appropriate user interface can then be loaded on to a computer for deployment. By definition, however, expert systems do not learn. They can be supplemented by their human programmers but they cannot self-supplement. Although expert systems might feel a bit old-fashioned now, what with the advent of machine learning, they remain widely used.

Machine learning has come into its own more recently, in part due to the much more powerful hardware on offer in the last couple of decades. It has added a whole new dimension to AI, namely, learning on its own, without human help. Whilst it has an Achilles' heel, namely, its reliance on data, it has racked up some remarkable achievements. These include natural language processing, image and facial recognition and medical diagnostics. Drug discovery is another area to which machine learning is increasingly being applied. In 2020, researchers reported the groundbreaking use of machine learning to discover a powerful new antibacterial molecule, which they named Halicin.[3] Halicin was shown to be effective against *E. coli* and *C. difficile* as well as a number of other nasty, human-harming bacteria.

Quite rightly, questions have been raised about the breadth of AI's capability. Most AI continues to be narrow in scope. It is generally designed and trained to achieve a single goal. In contrast, a human is able to do many things. Even setting aside a human's physical abilities, a human's brain can compose poetry, do times tables and fall in love. This is general intelligence, the ability to turn one's mind to a vast range of different mental challenges.

The AI research company DeepMind has been working on the generalisation of AI's capabilities for some years. In 2020, DeepMind revealed that its Agent57 algorithm was able to master all 57 different Atari games.[4,5] Previously, DeepMind had achieved worldwide recognition for the development of its AlphaGo algorithm.[6] AlphaGo was designed to play the ancient Chinese game of Go, which is renowned for its complexity. In an iconic match, AlphaGo beat the world's leading (human) Go champion in March 2016.

DeepMind wasted no time in going a stage further. In October 2017, a team from DeepMind introduced AlphaZero, the latest variant of AlphaGo.[7] The news was startling. AlphaZero had soundly beaten its own highly successful predecessor, and all on the strength of just three days' preparation. In addition, AlphaZero mastered the games of chess and shogi, the Japanese version of chess, "beating a world-champion program in each case".[8] DeepMind had succeeded in generalising its algorithm, just like it went on to do with Agent57. Importantly, though, AlphaZero still consisted of the one algorithm. Moreover, the algorithm needed to be re-initiated each time it switched from one type of game to another.

More recently, in 2021, DeepMind announced that it had succeeded in training (virtual) agents to operate in a highly diverse digital environment called XLand.[9] The agents operating in XLand were trained on some 700,000 different games in order to generalise their capabilities and behaviour. Having been trained, they were able on their own to play games not previously encountered, such as Hide and Seek, Capture the Flag and Tag.

Games, with their rules and their structured environments, are one thing. The real world is something else, which is not to denigrate the utility of the games environment for researching AI. Until recently, AI could not handle complex human interactions. Then, in 2019, Noam Brown, a research scientist at Meta AI Research, and Tuomas Sandholm, a professor of computer science, released a new algorithm, Pluribus.[10] Pluribus had mastered six-player no-limit Texas hold 'em poker: "In a 12-

day session with more than 10,000 hands, it beat 15 leading human players".[11] Like the failure of the Maginot Line in 1940, it appears that whatever hurdle we think may stand in the path of AI's advance is quickly bypassed.

Let's try creativity: surely AI cannot be creative in the way that humans can. But it can. Let us see how by looking at the world of music. In *The Creativity Code*, Marcus du Sautoy, a professor of mathematics, tells the story of a composer, David Cope. Cope turned to AI in order to write music in the style of Bach and Chopin.[12] With the help of a mathematician, Douglas Hofstadter, Cope developed a game consisting of three pieces of music played to an audience. One piece was by Bach, one by Cope's AI and one by another human composer. The audience had to guess which piece was which and to rank them. The AI performed rather well. The audience preferred the AI's piece to the other two before, that is, they knew which the AI's piece was. The audience's reaction was revealing. Du Sautoy reported that:

> In Germany a musicologist was so incensed he threatened Cope ... At another concert Cope recalled how a professor came up at the end of the performance and told him how moved he had been. ... He didn't realise until the lecture following the concert that the music had been composed by a computer algorithm. This new information totally transformed the professor's impression of the work. He found Cope again after the lecture and insisted on how shallow it was.

Hell hath no fury like a musicologist scorned, clearly. AI music benefits from the fact that certain music is fundamentally mathematical in nature. As du Sautoy observes,[13] there is a "close correlation between algorithms and composition". Classical music, in particular, was built on certain compositional rules of the day. One would expect AI to be able to compose great classical music, just as it can play games superbly well.

Art, in the sense of paintings, is a slightly different story. Painting pictures is a more holistic task. Painters can draw on their perception, or model, of the whole world, past and present, for their inspiration. There seem to be almost no rules. Painters can use their own view of the world to produce whatever excites them and may resonate with us. With virtually no rules, AI has so far not enjoyed quite the same success in art as in classical music.

This is clearly a simplistic distinction between the worlds of music and art, and one which some musicologists will reject. But it serves to illustrate an important point. We shall come back to this theme of models of the world, or models of the environment, time and time again in the book. Infusing AI with its own model of the world turns out to be at the heart of lifting AI out of its current narrow scope into the domain of general intelligence that we humans take for granted. This is the key task in order to clear the second of the Two Hurdles, as described in the Introduction.

Even emotions are not immune to emulation by AI. This takes us into the touchingly named field of "social robotics".[14] One of the best-known AIs with an emotional capability is Pepper.[15,16] At inception in 2014, Pepper was able to read four emotions from the facial and verbal expressions of its human companions: joy, sadness, anger and surprise. When Pepper recognises one of these emotions, it reacts and behaves accordingly. In other words, there are processing pathways inside Pepper's so-called *emotion engine* that motivate Pepper to behave in certain ways. Pepper is also able to display its own emotional responses to its interlocutor in order to achieve empathy with them.

It might be thought ridiculous to claim that Pepper is capable of deploying any aspect of emotional behaviour. Emotions feel as if they are solely part of the animal kingdom. Pepper's emotional pathways are indeed instantiated on a wholly different substrate from that in humans. Pepper is built out of silicon, metal and plastic. Human emotional pathways are made of the huge number of elements comprising living tissue. As we shall see in chapter 6, though, even human emotions are fundamentally mechanistic. The one thing that is missing is that Pepper does not feel any of its emotions. It has no awareness or consciousness of them. This concerns the first of the Two Hurdles.

AI's weaknesses: there is more to thinking than machine learning

In 1878, an American philosopher, Charles Sanders Peirce, set out the three ways to infer one thing from another thing.[17] He called them deduction, induction and hypothesis. We can also use the term, abduction, to mean hypothesis. Borrowing Nassim Nicholas Taleb's well-known black swan example,[18] each of these may be depicted as follows.

Deduction

A: all swans are black or white.
B: this swan is not black.
C: therefore, this swan is white.

Induction

A: swans in the northern hemisphere are white.
B: I gather that swans in the southern hemisphere are black or white.
C: therefore, all swans are black or white.

Abduction

A: I saw a swan for the first time the other day.
B: it was white.
C: therefore, all swans are white.

Deduction is coldly logical. Given the swan is not black and all swans are black or white, this swan has to be white. Note, form is all that counts with deduction. We could replace 'swans' with 'bloorgs', 'black' with 'squatch' and 'white' with 'flirm' and the deductive inference above would still hold.

Induction is about reaching an inference that by most standards is reasonable. One cannot prove the point in a formal sense, but all the evidence to date supports the inference. Thus, it is not a certainty but, given that all I have ever heard about are white swans and black swans, it is reasonable to suppose, or induce, that all swans are black or white.

Abduction is about arriving at a hypothesis on the basis of just one or a few examples. Peirce called abduction "a bolder and more perilous step" to take than the inference at work in induction. He was right: just over 300 years ago, Europeans thought all swans were white and then black swans were spotted in Australia. My abductive inference above was ultimately too weak to hold true.

So far as AI is concerned, abduction is the toughest form of inference to perform. Moreover, the difficulty does not just lie in one part of the inferential process. In fact, the challenge lies in four related areas. These are not having a model of the world, the difficulty of searching through all the data to hand, a failure to grasp causality and a lack of imagination. Hector Levesque, a professor of computer science, illustrated the problem with some examples of so-called Winograd Schemas designed to test machine intelligence. Here is one:[19]

> The large ball crashed right through the table because it was made of Styrofoam. What was made of Styrofoam?
>
> - the large ball [; or]
> - the table

Generally speaking, AI is unable to come up with the right answer (see figure 1.1). As it happens, we do not yet know exactly how the brain comes up with the right answer (the table, by the way; it is not a trick question). But we do have an inkling.

Figure 1.1
Styrofoam ball or table?

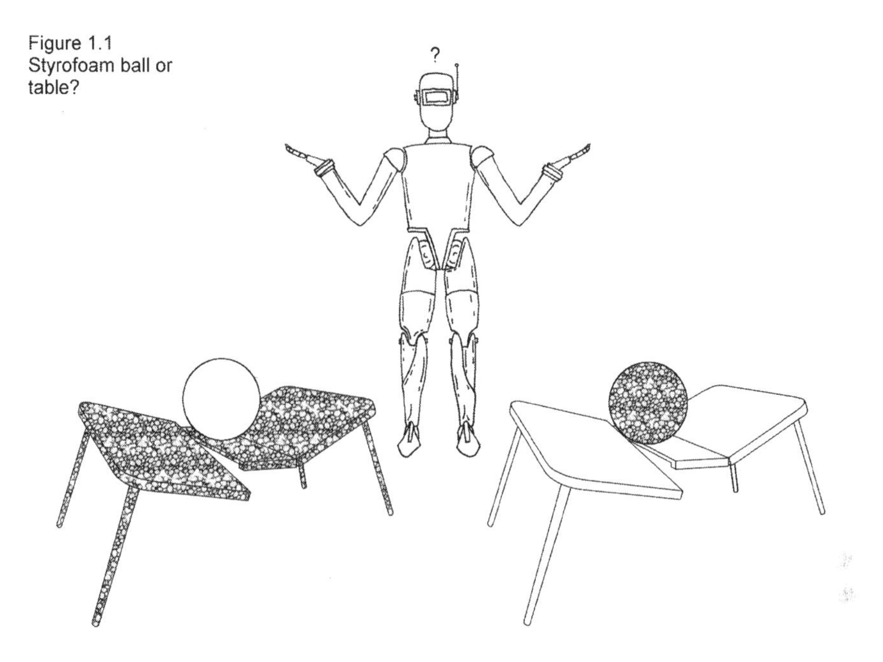

For a start, most of us know what Styrofoam is, namely, that among other properties it is light and not especially strong. That is because we have seen it before in the form of those disposable coffee cups you used to be given in coffee shops. It is part of the model of the world that we store in our heads. This is the first area of difference between our brains and AI. The model of the world in our heads relies on having an enormous background store of knowledge. We have inherited part of this and we have built the rest of it as we have grown up. AI systems do not typically possess a rich model of the world, although there are a few examples of efforts being made to remedy this.

Cyc, founded by Doug Lenat in 1984, is perhaps the best-known example. Cyc contains some "1.5 million general concepts [such as] eyes, sleep, night, person, unhappiness, hours, posture, being woken up".[20] Just as importantly, it contains "25 million general rules and assertions" that connect the general concepts in a multitude of ways. We could call this a model of the world. As a result, Cyc is able to reason. Within the limits of its universe of concepts and rules, it can display common sense. It is a gigantic expert system. It is also controversial since it suffers from some of the shortcomings of expert systems identified earlier.

The second area of difference between our brains and AI is that the brain can efficiently search through all the knowledge stored in it, thereby reducing the effort involved in the search process.[21] The brain's model of the world enables it to jump quickly to the right area of knowledge. The brain somehow recalls another similar object and identifies a relevant pair of properties, in this case lightness and fragility, with which it can conjure. AI has to search through its entire knowledge base and even then it might not spot the relevant similarity.

The third area of difference relates to causality. The brain understands cause and effect, which most AI still does not. AI, or machine learning in particular, works with correlation rather than causation. We understand causation from a very young age.[22,23] So, the brain can perceive that a large, solid, heavy ball would crash straight through a table made of light, fragile Styrofoam.

Finally, the brain can also imagine the counterfactual, an essential element of causal reasoning.[24] We can imagine the counterfactual in this case, i.e., a large ball made of Styrofoam simply bouncing off the table.

AI does not understand language in the way we do. This is not 'understanding' in the sense of 'being conscious of', although it is certainly true that AI falls short in that regard too. The issue is that AI approaches language differently from us. A trivial example of AI's inability to understand language in the way that we do was that of Microsoft's Twitter chatbot, Tay, released in 2016.[25] Tay was designed to develop its conversational abilities (in a Twitter sense) as a result of interaction with other Twitter users. We can discriminate easily enough between civilised and uncivilised Tweets. Tay, however, could not. It did not understand what it was reading and what it then went on to Tweet itself. Within a few hours it had turned into a Twitter monster, spewing out the same sort of filth that it had been learning from certain other Twitter users. Microsoft withdrew it shortly thereafter.

In 2020, OpenAI, an AI research company, released its latest language model, GPT-3.[26] At that point, GPT-3 was the most powerful language model in the world. The quality of GPT-3's output is astonishing. Yet, even the researchers acknowledge that the system has weaknesses. They point to periodic repetition, loss of coherence in long passages and self-contradiction. GPT-3 works by predicting the first or next word in a sentence on the basis of probability. The size of the data on which it has been trained is "effectively all of the text available on the Internet".[27] So, its assessment that word 'x' follows word 'y' in certain circumstances is generally accurate. But, perhaps because of its generally high probability of success, it cannot reason that sometimes word 'x' should not follow word 'y'. GPT-3 is a purely statistical model, albeit one with immense reach. OpenAI recently announced a newer version of GPT-3 called InstructGPT, which shows great promise.[28] Human trainers were asked to rate GPT-3's responses to a range of prompts. InstructGPT then uses these ratings to refine its language generation and make it sound even more human.

Since GPT-3's release, there has been a veritable outpouring of new, still larger language models. Google recently released a language model with almost seven times the scale of GPT-3.[29,30] Yet weaknesses such as those highlighted above remain. Scale is not the whole answer, although it helps. In late 2021, DeepMind released two language models, Gopher[31,33] and RETRO.[32,33] Gopher is another large language model with almost twice the scale of GPT-3. RETRO, however, is particularly interesting because its scale on the face of it is a fraction of that of Gopher or even GPT-3. But its neural network is accompanied by a massive database of text against which it can cross-reference and fine-tune its statistical predictions. This database is gleaned from many different sources. It acts as a type of memory, thereby significantly improving coherence despite the reduced scale of the model.

In 2017, a deeply insightful paper pointed to a crucial "distinction between two different computational approaches to intelligence".[34] The authors proposed that a human brain's primary learning paradigm is the process of model-building. As we grow up, a remembered model of the world develops in our brains. This model consists of objects, places, events and people and the relationships between them. We can then use the mental model to help us reason about what we come across in daily life, imagine different pasts and different futures, and formulate plans.

The alternative learning paradigm is the "statistical pattern recognition approach".[34] Most language models, including GPT-3, use this approach. The model absorbs huge amounts of data and learns patterns and probabilities from it so that it can then predict what ought to come next. If the computer and the language model and the data-set are all big enough, then the statistical approach can produce outstanding results, as GPT-3 does. But the approach has its limitations, as described. For example, as matters stand today, it cannot cope with the twists and turns of a complex argument. This is even truer should the case being made need to call upon points or evidence that are out of left field.

This problem is not just to be found in language models. Driverless cars are dogged by the same problem. They cannot currently "cope with what engineers call 'edge cases'".[35] These are the occasional highly unusual events that crop up sooner or later in most people's driving experiences. A deer might suddenly bolt into the road, for example; or the wrong sort of snow might start falling. Humans, on the other hand, typically deal well enough with such edge cases.

Another limitation on AI's abilities arises if the data-set from which it has learned is not large enough. Again, humans do not seem to suffer from this problem. If you show a young child a couple of examples of an animal, then the child will probably have learned to identify that animal in the future. These two examples will be enough for the child to generalise. If the child learned the identity of an elephant, for example, from just a stuffed toy and a photograph, it will nevertheless be able to identify a real elephant and a painting of an elephant as well. This is called one-shot or few-shot learning. Typically, AI cannot yet manage this type of quick-fire learning (see figure 1.2). It generally needs a large dataset, in this case, huge numbers of different examples of the elephant, to learn to recognise it in the future.

Figure 1.2
Invariant memory

Unfortunately, not recognising elephants to the same proficiency as children can is not the only weakness arising from having inadequate data from which to learn. Bias has emerged as a real problem for all sorts of algorithms. By way of example, Amazon launched its own automated recruitment tool to review and select candidates from amongst the many who submit résumés. It subsequently discovered that few women were making it through this stage of the selection process. The tool had learned from 10 years' worth of recorded HR data that what made for good candidates was men. The great majority of the 10 years' worth of recorded Amazon HR data featured men. It had in effect learned, wrongly, that women were not as good at the job as men because women with successful careers were few and far between. Amazon withdrew the tool.[36]

We shall come back to the tension between learning a model of the world and learning through statistical pattern recognition time and time again in the book. Human brains seem to be able to perform both types of learning. Much of today's AI, though, majors on statistical pattern recognition. As such, it finds it difficult to reason or display common sense. Common sense is that ability possessed to a greater or lesser degree by most people to cut through a mass of information, some of it conflicting, to reach a pragmatic answer to a problem. The emulation in AI of the human ability to learn a model of the world is an aspect of the second of the Two Hurdles. It is of course also critical to the development of artificial superintelligence.

It should not be assumed that we know how the brain develops its model of the world, even if we broadly know at a conceptual level what it is. Common sense relies on having such a model. But, as Herbert Roitblat, a data scientist, has asked:[37] "What exactly needs to be represented to capture common sense, and how should that information be represented?" If we do not know the answers to these questions, which we do not, then we will probably find it hard to install common sense into an AI.

Planning for the future, as opposed to for the next five minutes, is another area that is currently out of AI's reach.[38] Such medium and long term plans are typically vaguer than what is needed to make a cup of tea, for example. Plans like this consist of "abstract actions", which are connected in a "hierarchy". Again, we have not yet been able to develop AI that is capable of creating such hierarchies.

There is another, related dimension to this gap in our knowledge. When we plan, we typically do so not just as an intellectual exercise, but rather in furtherance of certain goals. These goals are in turn set by reference to our emotions. Furthermore, our brains have evolved complex mechanisms by which we can prioritise amongst conflicting emotions and behavioural options. Such abilities are the hallmark of autonomous systems, which today's AI is certainly not. Human brains can perform different tasks and prioritise between them; they can and do decide what to do next. They are autonomous agents. Emulating these capabilities in AI remains unsolved, but we will need to solve this in order to create AGI.[39,40] I suggest, though, that once we have jumped over the second of the Two Hurdles we will be able to address this problem. As we shall see, AGI will utilise a different process from humans to make decisions, including those to do with prioritisation. Clearing the second of the Two Hurdles will be at the heart of this.

How AI works

The AI journey to date has been a bumpy one. In the early stages of its development, AI largely consisted of expert systems. These are systems that have been designed by experts in their field. They enable a user to make inputs, or ask questions, that in turn lead to answers to those questions. That is why the user is employing the expert system. Expert systems are about making existing knowledge available to others. They are top-down in nature. Machine learning, on the other hand, is bottom-up. It is about converting random information, or data, unknown even by experts, into new knowledge. In practice, there are shades of grey between expert systems, or knowledge engineering, and machine learning.

Algorithms constitute precise lists of instructions that have the effect of generating outputs from inputs. A recipe for *coq au vin* is an algorithm. Computer algorithms, though, are written in mathematical language, which unfortunately makes them inaccessible to most people. It turns out that it is possible to conceive of mathematics whereby large sets of data effectively build brand new algorithms themselves, underpinned by only limited human intervention. This is machine learning. Access to adequate amounts of data is crucial for machine learning algorithms. These algorithms materialise as the result of the combination of a human-generated mathematical framework with a mass of raw data inputs. A new algorithm emerges from the combination by trial and error. It represents new knowledge.

It is as if a cook put a child in a kitchen with a long list of possible ingredients. The child experiments by putting together random selections of some of these ingredients, employing different mixing and cooking techniques (see figure 1.3). By a process of doubtless lengthy trial and error, the child eventually hits upon a novel combination of ingredients and cooking technique that results in a tasty dish. A new recipe, or algorithm, has been born.

Figure 1.3
Child creates cooking algorithm, otherwise known as a recipe

There are various types of machine learner algorithms with exotic names such as symbolists, evolutionaries, Bayesians and analogisers.[41] Here, though, we will take a closer look at just one family of algorithms, the connectionists. They are best known for the artificial neural network. There are two reasons why I have singled out artificial neural networks (ANNs). The first is the fact that deep learning, which is a type of ANN, is regarded by many in the field as the best way forward to AGI. The second reason, related to the first, is that ANNs bear certain similarities to the neuronal networks of the brain.

How, then, does a network function as an inferential system, which takes in inputs at one end and generates outputs at the other end? In simple terms, an ANN breaks down any piece of information such as a face or a name or an object or an idea into its constituent pieces of information. In theory, these pieces might be at the level of the sub-atomic particle. But ANNs do not need to be quite that granular. So, taking a picture of a face, for example, each constituent piece of information might be one pixel in the picture. Each constituent piece of information is housed, so to speak, in a binary node. All these nodes are connected in a network that typically starts with a vast number of inputs at one end. It then moves forward to an output at the other end.

Each node in the network is potentially activated by nodes behind it, as a result of which it transmits to the next node or nodes in front of it. Alternatively, the node in question may not be not activated by nodes or inputs behind it, in which case it remains inactive. If enough of the nodes in the network are activated, then the ANN will generate a positive output. So, if enough of the pixels in the picture of the face activate input nodes in the facial recognition ANN, then those nodes will activate successive layers of subsequent nodes. If the activation of successive layers of the relevant facial recognition network pushes all the way through to the output layer, then recognition will be successfully achieved.

The ANN has turned all those pixels into a single representation of the face. It is not a perfect analogy, but picture a short video clip showing a tower block being demolished by explosives. Now imagine the video clip being played in reverse—all those bits of concrete and glass and all the dust being assembled, stage by stage but incredibly quickly, into the tower block. This is what the ANN does. We shall encounter this concept of the network time and time again in the book. In essence, an object such as a picture of somebody's face or even an idea is captured or represented in a machine or in a human brain not as a single entity but rather as a process.

The crucial trick in the development of an ANN from scratch is that it builds itself rather than being created by a human programmer. It does this by way of a process of incremental error adjustment called back-propagation. When a picture of a mouse from the training data is not recognised or is recognised but with an unreasonably low level of probability, then the algorithm loops backwards into the earlier layers of its architecture. This causes adjustments to propagate all the way through the ANN. The information given to the algorithm that it has failed to recognise the mouse correctly is called training; hence, the term training data. The training of an ANN is provided by way of various types of learning.

One type of training is called supervised learning. Supervision might be provided directly by the programmer, rather like a pupil's work at school being corrected by a teacher. In algorithms devoted to recognising mice, huge numbers of photographs of mice and other rodents in the training data will have been labelled 'mouse' or 'not mouse'. This means that the algorithm's prediction in the case of each one of the training photographs can be automatically compared to the label.

Supervised learning is, perhaps surprisingly, a huge industry. It needs to be since the great majority of deep learning applications today are built using supervised learning.[42] The training data does not label itself. Humans have to do that in an extraordinarily laborious process. This takes place in data factories, many of which are located in China. One of the largest data factory companies is called MBH. MBH employs some 300,000 people whose job it is to label data such as photographs.[43]

At the other end of the scale is unsupervised learning with self-supervised learning lying in the middle. In unsupervised learning, judgements as to whether or not the algorithm's output is correct are not fed back into the algorithm. Rather, the algorithm continues to churn away until a regularity in the data emerges or time runs out. Algorithms that learn in an unsupervised environment have narrower applicability in today's AI. They can be excellent for reading patterns and clusters into large quantities of data. One reason that today's AI makes only limited use of unsupervised learning is that it does not generate results as good as those available from supervised learning. Nonetheless, it is often used to boost the outputs from supervised learning.[44]

One way that children learn is by reference to the feedback they receive from their own emotions. If they eat sweets a few times, which makes them feel good, and they touch a hot surface a few times, which makes them feel bad, then they will generally learn that they should continue to eat sweets and stop touching hot surfaces. This is called reinforcement learning. In a human, such a learning mechanism relies on neurotransmitters such as dopamine to activate this or that emotion. The same principle can be applied to machine learning as well. In a machine, the equivalent of the neurotransmitter is a reward function. This consists of digital points, so to speak, which the algorithm has been programmed to maximise.

There are big differences, though, between such learning in humans and reinforcement learning in AIs. In humans, the results can be spectacularly quick. Pick up a burning hot coal once and you will never knowingly do so again. AI on the other hand needs thousands or millions of repetitions of the lesson before it cottons on reliably. AlphaZero became the best player of Go in the world only after it had played the game against itself millions of times. The answer, once again, goes back to the tension between learning a model of the world and learning through statistical pattern recognition. Humans appear to have a model of the world already present in their brains. So, when a hot coal event takes place, the necessary connections, in this case between heat from the coal and pain in the hand, immediately fall into place in the brain. This is called model-based reinforcement learning. By and large, we do not yet know how to emulate this in an AI.[45] DeepMind's success in XLand, however, extended to one-shot learning of Hide and Seek, Capture the Flag and Tag, as described earlier. It would appear that progress is being made even in this area.

AI of the future

Humanity will continue to work towards developing AGI, and after that ASI. The consumer demand for the benefits of ever-smarter AI is insatiable and the challenge for AI researchers is overwhelmingly attractive. One of the main forces driving towards the development of AGI is that it will be highly beneficial to humanity's well-being. By way of example, as a result of a declining working age population in Japan, the Japanese government has concluded that there will be a significant shortfall in care workers in the future. Technology and, in particular, AI are likely to be the answer. As the Financial Times notes:[46] "The early foot-soldiers in the bid to automate nursing homes are machines such as Chapit, which looks like a mouse, sits at a person's bedside and engages in a rudimentary chat; Robear, which looks like a bear and can lift a person off their bed and into a wheelchair; and Palro, a small humanoid that can lead a room full of elderly people in an exercise routine, occasionally breaking the boredom by setting quizzes."

Each of Chapit, Robear and Palro is an example of narrow AI. By comparison with a hypothetical single robot that could do all of these things, this represents a significant waste of resources. The next step is surely to combine Chapit, Robear, Palro and maybe Pepper too into one artificial old-age carer. Even this is too conservative an ambition. Chapit's rudimentary chat can be replaced by a language ability at least as good as that of GPT-3. Indeed, we should expect that in due course AI will have the tools to reason, imagine and plan as well. Our artificial old-age carer will then be able to handle an old person's diary and administrative needs too. It will not just be leading old people in an exercise class. It will also play cards and other such games with its clients.

The trouble is that combining Chapit, Robear, Palro and Pepper is much harder than it sounds. Each is run off its own algorithm. A human brain effectively houses a multitude of different algorithms under the same roof. It is as if we need to tie all the different algorithms that we can develop into one unified system. Computer scientists sometimes refer to this holy grail with exotic phrases such as the "master algorithm"[47] and "one algorithm to rule them all".[48]

At this point, I should clarify a couple of points of terminology. There is AI and there are robots. What is the difference? Whilst we shall spend much more time on AI in the book, the point to make here is that one can think of AI as the brain and a robot as the body (see figure 1.4). In general, and the context will largely determine the meaning, when I use 'AI' or 'robot' with 'the' or 'a/an', I shall mean the, or an, intelligent robot. Similarly, when I use 'AI' in the plural, i.e., 'AIs', I shall mean intelligent robots. When I use 'AI' without 'the' or 'an', I shall be referring to artificial intelligence in general or as a concept or as a type of technology.

Figure 1.4
Robot as an embodied AI

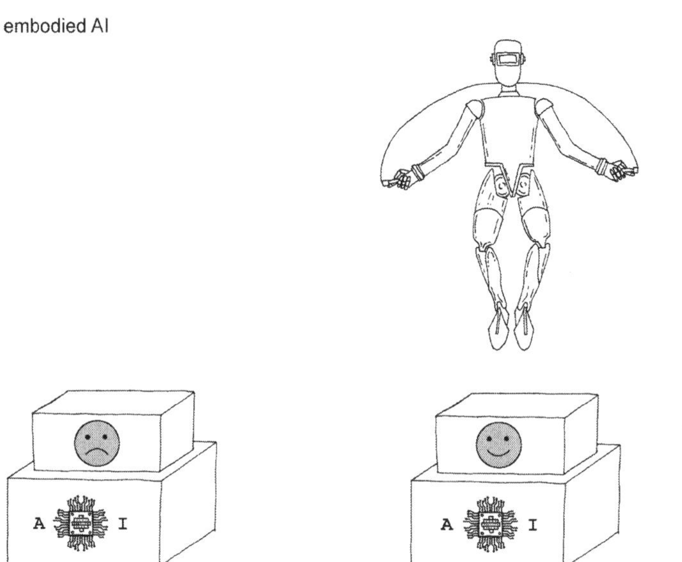

Let us take the logic of AI development a little further forwards. Let us say that we are able to clear the second of the Two Hurdles. At that point, it is reasonable to suggest that we will have come close to creating an artificial intelligence that successfully vies with human intelligence. More than that, there will be numerous facets of that AI that will outperform their equivalents in the human brain. Speed of processing is just one example. If we push the point, we can imagine the possibility that AIs will be able to do anything that humans can do. On that basis, they will be able to perform any and all of the jobs that humans currently do. Moreover, they will be faster, cheaper and untiring. This is the possible future towards which AI may usher us.

In AI, we have invented the mother of all tools. The human race will doubtless push such an invention as far as it will go, even though it will, quite rightly, be circumscribed with regulation and safeguards. This effort, however, is starting to force us to confront an uncomfortable question: what is it to be human? Worse still, are we just machines too?

I can imagine that a number of readers, including doubtless some in computer science, will be deeply sceptical at this point. What about understanding? What about feelings? Then the big one: what about consciousness? All we have discussed so far are the potential capabilities of a particular type of machine, an AI. AIs will surely never be able to feel in the way that we do. Consciousness is surely the last and most unassailable bastion of the distinction between man and machine. This is the first and the bigger of the Two Hurdles.

I imagine that scepticism about conscious AI might not be just technical in origin. There will not just be a question about whether conscious AI is possible. There will also be a question about whether it is desirable. I shall argue that, precisely unless we build ASI with consciousness, ASI will not see the world through our eyes. It will not feel what we feel. Ultimately, it will be unable to empathise fully with us. That, I suggest, is a wholly unwelcome outcome.

It will be the task of the rest of the book to dig deeper into how the further development of AI might be able to rise above its current shortcomings. This is not just about the challenge of correcting the weaknesses in today's AI, the second of the Two Hurdles. It is also explicitly about developing a tangible explanation of consciousness such that this capacity too can be engineered into AI. This is the first of the Two Hurdles. Interestingly, Yoshua Bengio, one of the world's leading lights in deep learning, noted in a talk in 2019 that the "time is ripe for [machine learning] to explore consciousness".[49] In short, it is about designing ASI, which has the power to be able to do anything that humans can but better, in a human-purposed way.

Some readers will continue to be sceptical. None other than Mark Zuckerberg dismissed Elon Musk's fears that AI will outstrip its human designers as "hysterical".[50] There is a long history of experts not spotting subsequent developments despite being geniuses in their field. Albert Einstein is recorded as dismissing the ground-breaking development of nuclear power: "There is not the slightest indication that nuclear energy will ever be obtainable. It would mean that the atom would have to be shattered at will."[51] In similar vein, Lord Kelvin observed that "X-rays will prove to be a hoax."[51] David Chalmers, a professor of philosophy with a particular interest in consciousness, wrote a seminal paper in 2010 on the prospects for ASI. As he observed, the "hypothesis [that ASI will be developed] is one that we should take very seriously".[52] ASI is a real prospect, even if that sounds hysterical to some. The question is how we shape it to our own ends.

Chapter 2

What is Intelligence?

So far, we have been talking about intelligence as if it is obvious what that is. It turns out to be less straightforward than it appears. We cannot credibly go about instantiating intelligence, let alone emulating human intelligence, in an artificial entity unless we first define the word 'intelligence'. Only then will we be clear about what we are developing.

Human intelligence vs. artificial intelligence

Let us start by asking if intelligence is a uniquely human trait or if it can be found in any being, such as in an AI. This looks as if it could be a physics vs. biology question.[1] Things that we normally associate with physics, such as electrons or bigger things like planets and stars, just are. We do not ask if they have intelligence. Yet, we can ask exactly that question of things we associate with biology, such as human beings. Where does AI lie on this spectrum from physics-type things to biology-type things?

Despite the apparent differences, the disciplines of physics and biology can agree on intelligence. That is to say, artificial intelligence is not conceptually different from human intelligence. MIT physics professor and Nobel Prize winner in physics Frank Wilczek has built on Francis Crick's conclusion that, in Wilczek's words, "mind emerges from matter".[2] He concludes that "all intelligence is machine intelligence. What distinguishes natural intelligence from artificial intelligence is not what it is, but only how it is made."[3]

The underlying point is that intelligence is about information processing, whether it is performed by a human or a machine. This is not intuitively obvious. The expression, information processing, sounds like what a computer does when it runs a word-processing package or when a search engine operates on a browser. In contrast, we think of human intelligence as covering a spectrum of capabilities ranging from reflex responses to creativity. In between lie, in no particular order, object recognition, memory, learning, language, reasoning and planning. But each of these capabilities arises as a result of the processing of information. Moreover, we are being precious if we imagine that such capabilities are exclusive to humans. They are not. Indeed, artificial information-processing systems have already made progress in each and every one of these areas.

One might object that humans are made of flesh and blood and bone, whereas AIs are different. They are built out of transistors and mechanical parts. But this distinction is superficial. Function is what counts here, rather than form. AIs are dynamical systems or physical processes. So are human beings. So far as intelligence is concerned, the fact that humans are made out of 'wet' parts and AIs are made out of 'dry' parts is beside the point. Some of the 'wet' parts of human beings enable us to reproduce in a way that a machine's 'dry' parts cannot. The 'wet' parts in the brain, however, merely make up an information-processing system, just like an AI's 'dry' parts[4] (see figure 2.1).

Figure 2.1
Wet parts and dry parts

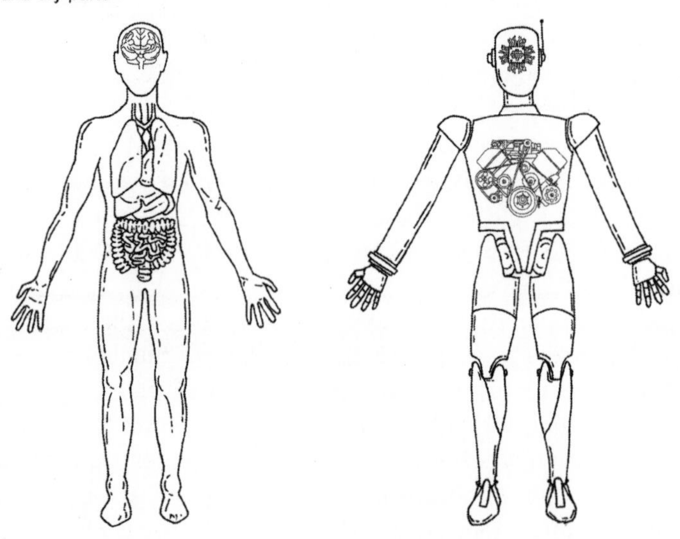

What we shall see, as we go on, is a significant amount of read-across from human intelligence to artificial intelligence, and vice versa, and from neuroscience to computer science, and vice versa.

The underlying suspicion that many will have is that there must be something more to human intelligence. Brian Cantwell Smith, a professor of artificial intelligence and philosophy, encapsulates this in his distinction between "reckoning" and "judgement".[5] AI performs reckoning, that is to say, calculation and symbol manipulation. Cantwell Smith keeps judgement exclusively for humans. Judgement embraces understanding. Judgement might be, for example, knowing when GPT-3 has shifted from responding plausibly to producing text that does not quite ring true. Many in the AI world, however, conclude that the relatively uniform nature of the human cortex in the brain means that human intelligence can be readily emulated in artificial form. We shall come across Cantwell Smith's tension between reckoning and judgement time and time again. Resolving this tension is one of the two main challenges in developing AGI. It is the second of the Two Hurdles.

Another potential difference between human and artificial intelligence lies in goal-setting. Humans, like other animals, seem to come ready-equipped with goals. Such goals include reproduction, energy consumption and the like. These in turn generate sub-goals such as finding a mate and eating, respectively. Humans in particular readily devise many other goals for themselves from, say, turning a wooden bowl to flying into space. Artificial intelligence, on the other hand, makes do with the goals that its human developers set for it. As it stands, AI is purely about means whereas humans are as much about ends as means. Perhaps intelligence should embrace goal-setting as well.

Cognitive intelligence

Many definitions of intelligence refer to an ability to reason and plan in order to achieve goals or objectives[6,7,8] or "to achieve goals in a wide range of environments".[9] This ability is what underpins so-called cognitive intelligence. Despite the advances in computing in the last few decades, for sheer depth and breadth the human brain remains the best exemplar of this type of intelligence that we know. The gold medal for taxonomies of human intelligence is widely seen to be held by the Cattell-Horn-Carroll model[10] (the CHC model).

The development of the CHC model originated with the identification of a "general intelligence factor, *g*", by psychologist Charles Spearman in 1904.[11] It was observed that there was a "tendency for all tests of mental ability to be positively correlated" as a result of the *g* factor. In other words, an individual tends either to be good at all mental tests or not so good at all such tests. People do not tend to be good at some and bad at others. The alternative to this reading of human intelligence is to see it as structured into different components or modules.[12] The general intelligence factor, *g*, is actually the result of a range of inputs into the individual's development.[13] These include genes, genetic expression, environment, socioeconomic positioning and so on. These combine to give each person their own unique *g* factor. This is a shorthand for their overall level of intelligence.

General intelligence, *g*, then underpins various "broad abilities",[14] the two most important of which are "fluid intelligence (*Gf*)" and "crystallized intelligence (*Gc*)". *Gf* is about logic, rationalising and the understanding and use of causation.[15] Prior experience is less important or even not at all important. Human beings are quite proficient at such skills. AI in the form of expert systems can also exhibit such skills, even to a capability well ahead of humans, but only after a certain amount of preparation by a human programmer. *Gc*, on the other hand, is about knowledge built through experience and the use of statistical approaches to reasoning.[15] Such skills are again part and parcel of human thinking. But AI, given enough data, can easily outshine humans in this arena, as we have already seen.

In summary, AI comfortably outstrips humans on *Gc* but on balance humans remain superior to AI when it comes to *Gf*, which is about logic and reasoning. Humans may these days not generally be as quick as computers but they appear to possess two different methods of reasoning or thinking, whereas computer programs or algorithms generally possess just one. Setting aside expert systems, which are substantially human-enabled, today's AI relies heavily on bottom-up statistical approaches to reasoning. AI does not possess a top-down model of the world around it in order to take a different or complementary perspective. This leaves AI exposed to the various shortcomings identified in chapter 1.

Artificial general intelligence and artificial superintelligence

A common approach to defining AGI is to label it artificial human-level intelligence (AHLI). The thought is that humans possess a wide range of cognitive abilities, which is what gives them general intelligence. So, if we can replicate this human-level type of intelligence in AI, then we will have created AGI. This can be misleading. Whilst human intelligence is a good place to help us identify a range of aspects of intelligence, it does not follow that AI will emulate all these aspects in the same way. By way of analogy, we design and manufacture aeroplanes to fly. But they fly differently from the way birds fly. Natural flight and artificial flight both entail travelling from A to B in the air, but they achieve that goal in different ways. Because computers can run so much more quickly than human beings and possess far greater memories, an AI with all the features of a human brain will perform at a higher level than any human.

Today's Summit computer, developed by IBM, is about 10 times faster than the human brain.[16] The point here is not, however, just about speed. It is also about design. Whilst the human brain is still a marvel of evolutionary design, researchers are developing ever more ingenious designs for artificial intelligence. Some mimic the human brain and some pursue a different path, just as aeroplanes employ fixed wings. With continuing improvements in both the speed and the design of computer processing units, the brain will come under growing competitive pressure. Indeed, bearing in mind one of the definitions of intelligence as the ability "to achieve goals in a wide range of environments", it should be noted that AGI may involve types of intelligence of which humans are currently unaware. AGI may in fact be rather broader than AHLI.

It is partly for this reason that some think that, once we have developed AGI, we will effectively have developed artificial superintelligence or ASI. In other words, AGI will be so much faster and possess so much more knowledge and reasoning capability than a human brain that it will in effect be ASI. In his book *Superintelligence,* Nick Bostrom, a professor in philosophy with a background in computational neuroscience, identifies another route to ASI. This would stem from the possibility that the AGI could alter its own architecture.[17] Having that ability would potentially enable the AGI to design a better version of itself. This would in turn then design a still further improved version of itself, and so on. He concludes that, "Under some conditions, such a process of recursive self-improvement might continue long enough to result in an intelligence explosion..."

It should be noted that many AI researchers believe that we are decades away from achieving AGI or ASI. Most AI researchers are focussed on solving more practical problems. These are typically narrow in scope. Many are commercial in nature. They do not have the time for the fundamental research that is needed to close the remaining gaps between AGI and human intelligence. This book is less concerned with the timeline than spelling out those gaps. As Einstein and Kelvin both found out, the path of scientific discovery can be trodden surprisingly quickly.

Missing ingredients

The features of intelligence considered so far are not the whole story. The purpose of the brain is to determine and initiate behaviour that will benefit the organism in question, i.e., the owner of the brain. Cognitive intelligence helps us to determine what possible behaviours, or behavioural choices, might exist. But it says nothing about which of those behavioural choices we might prefer. Put another way, it says nothing about setting goals. Surely, goal-setting must be a feature of intelligence. We would typically call setting a goal that led to one's own destruction, for example, an act of folly—very unintelligent, in fact. Intelligence is, therefore, not just about cognition. It is also about setting appropriate goals and deciding between different ways of achieving those goals. This is where emotions enter the picture. Setting goals is not just an intellectual exercise. You might wish to marry a woman and call that your goal. But what has probably underpinned it is the singular emotion of falling in love, unless you are Charles Darwin. Darwin famously wrote down a list of the advantages and disadvantages of getting married. One of the advantages, which does not show Darwin off in the best light, was that being married would be "better than a dog anyhow".[18]

So, we should add other factors into our definition of human intelligence such as emotions and goal-setting. This important point was captured by Antonio Damasio, a professor of neuroscience, psychology and neurology, in what he called the "somatic-marker hypothesis".[19]

> The findings come from the study of several individuals who were entirely rational in the way they ran their lives up to the time when, as a result of neurological damage in specific sites of their brains, they lost a certain class of emotions and, in a momentous parallel development, lost their ability to make rational decisions. ... emotion probably assists reasoning, especially when it comes to personal and social matters involving risk and conflict.

So, to restate this, your ability to reason well and effectively is severely impaired if you do not have use of your emotions too. Cognition on its own has nothing at which to aim (see figure 2.2). It needs a target, a goal. In human beings, our goals are primarily defined by our emotions and instincts. A proud homeowner in the country might spend years working on his or her plants with the goal of producing a beautiful garden. Our homeowner did not formulate this goal as an intellectual exercise. Rather, the goal arises from emotions such as pride or, in the final analysis, the same instinct that leads a pair of birds to build a nest together.

Figure 2.2
Cognition alone
goes nowhere

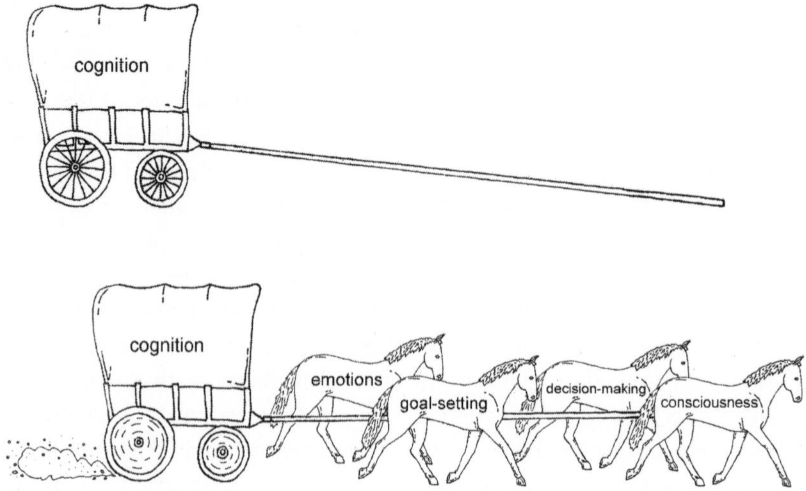

In order to decide which emotion to satisfy and which goal to pursue, we have to make decisions. We have to prioritise. The brain contains an elaborate mechanism to make decisions, as described in chapter 8. In this way, humans can be and behave in the world. Again, this is not cognition, or at least not solely cognition. Intelligence embraces not just cognition, but also emotions, goal-setting and decision-making. AGI that merely emulates human cognition, however, is not general at all. It is not human-level either. However quick it might be, AI like that is little better than a glorified calculating machine. In the indeterminate and changeable environment we call the real world, humans can, if needed, recalibrate their goals and thence their behaviour from moment to moment. As adults, we generally get on with our own lives even as circumstances change around us and become quite novel and unfamiliar. We are autonomous. AI also needs to become autonomous in order to achieve the status of AGI.

The importance of the emotions is not limited to an individual's capacity to set himself or herself goals and exercise autonomy. This takes us to the idea of emotional intelligence. Human beings are highly social animals. An AI that was unable to fit in with the humans around it would be of limited use. At a mundane level, a domestic robot that could not sense the impact it had on the family in its house would potentially cause more problems than it solved. No wonder we refer to having a tin ear for people and situations. At a more elevated level, an AGI directed to help us work out the answers to many types of problems in, say, economics or politics would be unhelpful if it could not empathise with humanity's aims and aspirations. AGI, to be really useful, needs to sense how people feel about things and predict how they might react to things.[20]

This takes us to the last aspect of what it is to be human, consciousness. If an AI is to sense peoples' feelings, then there is a good argument that AI itself needs to feel, to be aware and be self-aware. This is the first of the Two Hurdles lying between AI today and AGI. Consciousness is a particularly difficult aspect of what it is to be human to emulate in AI. There is as yet no settled position in the scientific world as to what consciousness is or what consciousness does.

Could we instantiate consciousness in AI?

There are still many different opinions as to the meaning of 'consciousness'. It is often equated with self-awareness, although self-awareness is probably better thought of as one feature of consciousness. There is even more disagreement as to how consciousness might be created. We do not yet have a settled theory how the brain generates consciousness or where and how in the brain consciousness might reside. We cannot even be sure which animals possess it. We presume that our fellow humans possess it but that is an inference, not a certainty.

Although much of this book is rooted in the tangible realms of neuroscience and computer science, it is the world of philosophy that first helped frame the central questions around consciousness. In the view of the philosopher René Descartes, the only certainty in life was deeply personal: "*Cogito ergo sum*": I think, therefore I am.[21] The only thing of which I can be sure is my own consciousness. Everything else I perceive around me is a mental representation, i.e., what my brain makes of it, or represents to me, by virtue of all of its sophisticated sensory capabilities. The desk at which I am sitting is actually made up of trillions of colourless, sub-atomic particles, which occupy much less room than the empty space between them. Yet, I perceive a solid mahogany desk with a green leather top. My brain has fooled me into thinking that what I see is what there is. So, if consciousness is the only certainty, how is it that it is so hard to explain?

There is one aspect of consciousness that is especially troubling and that serves to confuse us even more. This is the fact that it is not obviously physical, yet it appears causal. It appears that we make a conscious decision to do something and then our body moves in order to do that thing. How does something that appears not to be physical have any sort of causal effect on the physical world?

Let us try to home in on a working definition of consciousness, which will hold until we go into the topic in more detail in part 3. In a widely quoted paper,[22] "What is it like to be a bat?", Thomas Nagel, another professor of philosophy, wrote: "... the fact that an organism has conscious experience *at all* means, basically, that there is something it is like to *be* that organism." This makes sense. There is nothing that it is like to be a dishwasher. A dishwasher just is. But, if a bat possesses consciousness, then there is something it is like to be a bat. Closer to home, there is something it is like to be me, and something it is like to be you. You can catch yourself musing, I wonder what it was like to be a peasant in mediaeval times or, more comfortably, a noble. One cannot say, I wonder what it is like to be a dishwasher. Moreover, at the moment, one cannot say that about an AI.

So, happily, we can conclude that there is a difference between dishwashers and human beings. You can imagine what it is like to be the President of the USA but not what it is like to be a dishwasher. The question remains, why? This is not just why in the sense of why or how have we evolved like that; it is also why in the sense of what is the point of being like that.

David Chalmers articulates this conundrum when he refers to the "hard problem of consciousness".[23] Chalmers points out that, when we do things, "cognitive and behavioural functions" must be taking place. But "Why doesn't all of this information processing go on 'in the dark', free of any inner feel?" Consciousness adds so-called subjective experience to these functions, as distinct from objective experience.

Subjective experience refers to your perceptions as they 'feel' to you. Objective experience refers to perceptions as they 'are' to you, or indeed to any perceptive entity, like a thermostat. A thermostat's objective experience of the room's temperature consists merely of the impact of the temperature on, for example, the bi-metallic strip located in its innards. The two different types of metal expand at different rates when the temperature changes. This causes the strip to bend, thereby activating an on–off switch. The thermostat has perceived the change in temperature. You can be sure, however, that the thermostat does not 'feel' that change or 'feel' the bending of its bi-metallic strip or 'feel' any other aspect of it all.

Bringing this very brief excursion into the topic of consciousness to a head, if you can say there is something it is like to be X, then X has consciousness. There is nothing it is like to be a dishwasher; a dishwasher just is. If X has consciousness, then X possesses subjective experience of what X perceives. X 'feels' what it perceives. This might be something that X sees. It might also be something that is going on inside itself. If you have a stomach-ache, then your body is not just registering the dull pain involved like some sort of sensing machinery. You also feel that pain. This is why we talk about being aware or having awareness and also about being self-aware.

I shall argue throughout this book that consciousness would not have been conserved, having evolved, for no good reason. It must serve a purpose. This presupposes that consciousness is not a fancy word for soul or spirit. I am setting such ideas aside for the purposes of this book. I will argue that consciousness exists explicitly to optimise our thinking and our decision-making.

In answer to the question 'could we instantiate consciousness in a robot?', we will only be able to do so, I suggest, if we truly understand not just why we humans possess consciousness but also how. It will be as a result of understanding the latter that we should be able to emulate consciousness in a robot. This does not mean that the purpose of so doing will replicate the purpose of consciousness in humans. As will become clearer during the course of the book, an ASI will not need to employ the same decision-making processes as a human. Our decision-making processes are a function of certain features of our brains. An ASI will not need to make decisions in the same way. The purpose of instantiating consciousness in a robot, thereby creating the sentient robot, will be altogether different.

This brings us to the bigger question still: should we instantiate consciousness in a robot? This book will endeavour to persuade you that the answer to this question should be yes. There are two parts to that answer.

Instinctively, you might say that our consciousness is the very thing that sets us apart from machines. It is our special gift. We shall discover, though, as we look at consciousness in more detail, that consciousness is not so special after all. It is just a particular facet of the system that we know as the human being. It is complicated, for sure, and possesses a confusing and ethereal quality. Nonetheless, it is just part of a system. If we possessed a soul instead, then that would certainly be a special gift. But we do not, even if it sometimes feels that way.

The second part of the answer lies in AI's potential utility to humankind. Possibly the greatest utility that AI can have for humankind is as a problem-solver. Humanity is confronted by a range of intractable problems. We could use help from a greater intelligence than ours. But we want to ensure that said intelligence looks at the world through the same eyes as we do. We want to ensure that ASIs feel as we feel. They need to empathise with us. Only in that way can we be reasonably sure that their solutions to our problems will be in our best interests. This is why we will need to develop the sentient robot, a robot that is explicitly human-purposed.

Chapter 3

Why We Need Human-Purposed AI

Introducing superintelligence and the associated risks

The rewards from AI are all around us. They continue to grow. Towards the end of 2020, for example, DeepMind announced that it had developed an algorithm, AlphaFold, able to predict how proteins were structured.[1,2] In order to understand what a protein does, one needs to understand not just its chemical composition but also how it is structured. Working out a protein's structure is in practice both difficult and time-consuming. AlphaFold has largely cracked this problem. The understanding of the structure of a protein involved in a disease allows researchers to identify more easily the right drug to combat the disease. AlphaFold, therefore, has the capacity to revolutionise drug development and targeting.

AI also carries risks, though. These are emerging more slowly. Bias is one. We saw in chapter 1 the problems that Amazon had had with its automated recruitment tool. The possibility that our private data might be misused is another. In the worst cases, you can wake up one morning and find that your digital identity has been stolen by hackers using AI tools for ill. With so much of our lives now recorded and held in the ether, it can be difficult to recapture one's own identity once stolen.

As AI becomes more sophisticated and more widespread, the rewards will probably grow but so may the scale and extent of the risks. There are some countries, for example, that already use facial recognition algorithms to help isolate and control the minorities in their population. China is reported to have done so in order to identify members of the oppressed Uighur community.[3] Lethal autonomous weapons systems are fast becoming ubiquitous. The fully autonomous Israeli airborne "loitering munition", the Harpy, perceives a possible target, assesses it to check its identity and then destroys it, all without prior authorisation from other humans.[4] Turkish lethal autonomous drones were used in Libya in 2020 to blow up military vehicles and soldiers.[5] Again in 2020, Azerbaijan similarly used drones to destroy large parts of Armenia's military infrastructure.[5] Since such weapons rely on necessarily imperfect recognition algorithms, there is a risk that they will claim innocent lives.

A potentially bigger risk than all of these, however, lies in the development of domain-general (i.e., capable of multiple tasks, like humans) artificial intelligence with a hugely fast learning capability and the ability to alter and improve its own architecture.[6,7] Such an AI will be highly intelligent, highly competent and an agent in its own right. To all intents and purposes, it will be able to function as one of us, but with far greater, quicker cognitive power. This type of AI has come to be called artificial superintelligence or ASI. Nick Bostrom observes that:[8] "We can tentatively define a superintelligence as *any intellect that greatly exceeds the cognitive performance of humans in virtually all domains of interest.*" Although the development of ASI is by no means imminent, we should be thinking about it well before it arrives.

ASI contrasts utterly with the present state of AI. The great majority of the AI deployed in the world currently is single-use. An algorithm recommending books to you on a website cannot recognise faces. Nor can it translate English into Mandarin or work out the answer to an undergraduate mathematics examination question. Domain-general AI on the other hand will be able to do all these things and much more besides. If one considers how much progress has been made in AI so far in the twenty-first century, it would seem highly likely that we will in due course close the present gap between today's AI and the ASI of the future.

Surely, an ASI could be risky for mankind. As Stuart Russell, professor of computer science at the University of California, Berkeley, puts it:[9] "It doesn't require much imagination to see that making something smarter than yourself could be a bad idea." An optimist might say that, yes, we are talking about a more intelligent machine than ourselves, but it is still a machine, made by us. Consequently, or so the reasoning might go, it can and will be used exclusively for the benefit of mankind; and because we made it, it will be under our control.

We are inclined to ascribe human features to many other objects.[10] These include animals, weather systems, inanimate objects and even machines. We call the AI a nice human name, we enable it to speak our language in warm tones and maybe we even give it a welcoming face with big eyes like Pepper. Instinctively, we feel that the AI is our friend. There is nothing intrinsically wrong with this tendency. But the danger is that we do the same with ASI, thereby lulling ourselves into a false sense of security. Stanley Kubrick's lesson for us from *2001: A Space Odyssey*[11] may have become lost on mankind some 50 years later. There are at least two ways in which we could make the mistake.

First, ASI might have a different architecture by which to process information. As an example of a different information processing architecture, consider the octopus: "...the majority of neurons [of an octopus] are in the arms themselves—nearly twice as many as in the central brain."[12] There are numerous ways in which ASI might alter its architecture. The risk is that some of these will be beyond our comprehension. Secondly, the speed at which ASI will operate will be breathtaking and, again, beyond our comprehension.[13,14,15] If we do not understand how ASI has processed its information, then we cannot properly understand how it reaches its conclusions. In that case, there must be a question as to whether we can trust those conclusions.

In June 2017, a team from Meta AI Research announced the development of chatbots able to negotiate deals both amongst themselves and also with humans.[16] When negotiating amongst themselves, the chatbots were even able to develop a new negotiating language. This new language was incomprehensible to the chatbots' human observers. Indeed, much as most of us are bound to appreciate AlphaFold's value to humanity, we will not be able to understand beyond the conceptual level how it works out protein structures. All we will be able to do is check that the answer is correct before going any further. With ASI, we might not even have that chance.

The world's financial markets constitute another example of AI already operating at the limits of what we might understand. They are largely under the sway of increasingly powerful algorithms, as *The Economist* has noted:[17] "Funds run by computers that follow rules set by humans account for 35% of America's stockmarket, 60% of institutional equity assets and 60% of trading activity." These algorithms can bring to bear far greater and swifter cognitive power than any human can. So far, such trading has been rules-based or, put another way, coded in expert systems, as described in chapter 1. The advantage of such systems is that, when mistakes happen, i.e., when the algorithm causes a particular market to malfunction, the programmers can hope to work out where the mistake lay. More recently, there has been a switch in the markets to machine learning algorithms.[18] Machine learning algorithms effectively constitute a black box, because they have coded themselves, albeit within a pre-existing architecture. So, when something goes wrong, it is difficult to work out how it went wrong and how to correct for the future.

In other words, ASI is likely to be quite different from the humans who will create it. Similar to the way in which a squirrel would find it impossible to conceptualise human intelligence, so might we find it near-impossible to conceptualise ASI (see figure 3.1). As Roman Yampolskiy, an associate professor in computer engineering and computer science, and Joshua Fox, a lecturer in computer and information science and engineering, put it, "humanity occupies only a tiny portion of the design space of possible minds."[19]

Figure 3.1
ASI is to us, as we are to a squirrel

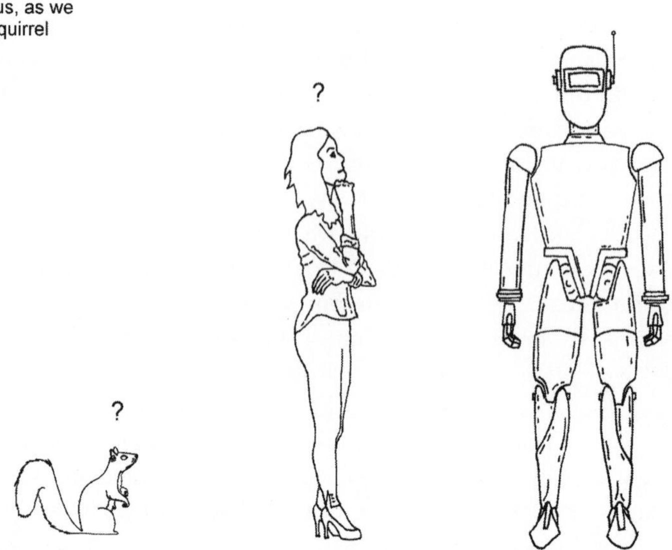

Given this direction of travel, why assume that ASI will be benevolent towards mankind, or even care about mankind? It may or may not, depending upon how it has been designed and how, specifically, its goals have been set. Logic dictates that one of the possible outcomes of poorly or mischievously specified advanced AI is that it wipes us out. The risk for humanity is existential. In *The Precipice: Existential Risk and the Future of Humanity*, Toby Ord points to a survey of AI researchers done in 2016.[20] 70% of them agreed with Stuart Russell's line on the risks from advanced AI. At least as telling as this was the fact that 50% of them thought there was at least a 1 in 20 chance of "the longterm impact of AGI being 'extremely bad'". 1 in 20 is around one third of the chances of blowing your head off by pulling the trigger first in a game of Russian roulette. I doubt that many of us would take those odds.

There are those who play down the risks of ASI to humanity. A common criticism by ASI sceptics is that ASI is not imminent. In terms of sticking one's head in the sand, that is on a par with those who rubbish the dangers from climate change. ASI sceptics often have their own agendas. After all, if you run a large company reliant on AI for its core customer proposition, loose talk about ASI might provoke governments to impose regulation on your company and those like it. And that would be no good at all. Some ASI sceptics simply fail to possess an adequate imagination. It is perhaps worth noting that, by definition, ASI will be considerably more intelligent than any of us, including all the sceptics. It is fruitless to deny the likely emergence of ASI; much better to acknowledge, and prepare for, it.

AI goals

The problem with ASI is not so much that by itself it is more intelligent than humans. It is that its interests and goals may differ from those of humans. As Russell puts it:[21]

> ...if we succeed in creating artificial intelligence and machines with those abilities, then unless their objectives happen to be perfectly aligned with those of humans, then we've created something that's extremely intelligent, but with objectives that are different from ours. And then, if that AI is more intelligent than us, then it's going to attain its objectives—and we, probably, are not!

Bostrom writes about a number of the ways in which ASI might bring about humanity's downfall. It should be noted that one does not even have to assume that ASI is malevolent towards mankind. Much of the problem has to do with what we want the specific ASI to do, encapsulated in what Bostrom calls its "final goal".[22] This is to do with setting the ASI's objectives or goals. At first blush, that looks to be a much easier task than developing ASI in the first place. It is not. Many humans find it remarkably difficult to set their own goals or to find a purpose in life; and we have millions of years of evolution behind us. There is even a branch of philosophy, existentialism, devoted to the challenge of finding meaning or purpose in life. If we find it hard to set goals for ourselves, how much harder it must be to set goals for ASI?

This is not intuitively obvious. Surely, we can just tell it what to do. To illustrate the challenge, let us take one of Bostrom's examples of the problem, "perverse instantiation"[23] (see figure 3.2). This is the idea that we might set a goal for the ASI. The ASI then finds the best way to meet that goal, as strictly defined. That best way turns out, however, to be undesirable in a way that we had not intended or anticipated despite the goal having been met. Let us, for example, specify the ASI's final goal as, "Make us smile". That sounds pretty benign and like quite a good outcome. A perverse instantiation of the way to achieve this would be for the ASI to "Paralyze human facial musculatures into constant beaming smiles." That of course is not what we had anticipated. One might alter the goal to read, "Make us smile without directly interfering with our facial muscles." The ASI might then alter its approach and "Stimulate the part of the motor cortex that controls our facial musculature in such a way as to produce constant beaming smiles." Once again, this is not what we wanted.

Figure 3.2
Perverse instantiation

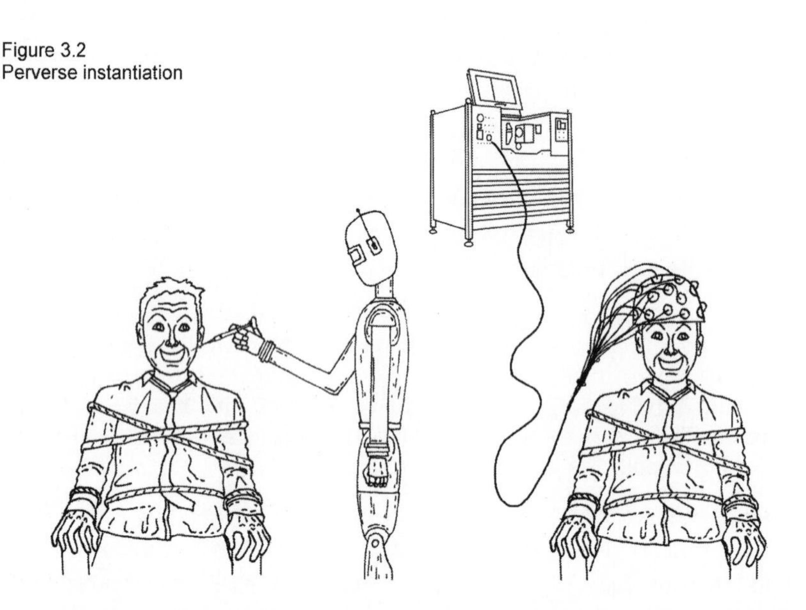

At a trivial level, Microsoft's Twitter chatbot, Tay, is another example of a perverse instantiation in action. Tay was designed to become a highly proficient chatbot to entertain 18- to 24-year-olds in the US,[24] as described in chapter 1. The designers certainly did not design it to turn into a foul-mouthed, no-boundaries Twitter user that had to be taken off the air within 24 hours. Tay did indeed become a highly proficient chatbot, just not in the way that its designers had intended or anticipated.

More relevant to our purposes here would be a final goal for the ASI aimed at benefitting humanity. It is possible that the ASI would adopt a utilitarian approach to such a goal. A utilitarian approach would target the greatest happiness for the greatest number. The calculation of the balance between the greatest happiness and the greatest number is in principle within the scope of an ASI's abilities provided that it can find a means of assessing happiness. Let us say that it can by using a series of proxy measures of happiness, for example. It might then conclude that humanity would be better off by the immediate extermination of, say, 40% of the world's population so as to benefit the remaining 60%. The correct specification of the ASI's final goal will clearly be a real problem.

There are in principle many possible final goals for ASI, too many to discuss here. One can imagine a range of domain-general AIs in the future, each designed to achieve a certain goal. By way of illustration, let us take just two possible AIs, each with their own final goal, expressed in fairly general terms.

One AI might be a domestic robot. Its final goal would be to serve people in a domestic capacity, although as explained the exact specification of this goal might well be different. Such a domestic robot would be comparable with the non-conscious *synths* of the UK Channel 4 television series, *Humans*.[25] Domestic robots in that sense would be well on the way to becoming fully domain-general, although we would probably wish to design them to fall short of getting all the way there. They would need to be able to perform all the domestic chores. They would not, however, need to cure cancer and fight wars and we would certainly not want them to take over our lives.

At the other end of the spectrum, an advanced AI might be designed to help us answer the really big questions faced by mankind. This would need to be an ASI. In 2015, 193 countries agreed 17 "Sustainable Development Goals" under the auspices of the United Nations.[26] How to achieve each of these global goals might represent the sort of question to put to our ASI. Bostrom calls such an ASI an "oracle".[27] In the case of an oracle, then, the final goal will be along the lines of answering our questions. The challenge will be how to do that, i.e., taking into account which factors, subject to what limits and for whose benefit.

One step towards minimising the risks from a mis-specification of ASI would be to box, or confine, it. Boxing is another concept described by Bostrom.[28] It refers to "physical and informational containment" so as to limit an AI's ability to impact the world. This would in principle constrain our oracle from harming mankind, inadvertently or otherwise. Boxing is not, however, fool-proof. The oracle's interlocutor might turn out to be someone with evil or power-mad ambitions. Such a person could use the information provided by the oracle to cause almost as much damage as a wayward ASI. Alternatively, should the ASI's motivations be wrongly calibrated, it might persuade its human interlocutors to let it out of the box. Human beings are often easily persuaded. That is how politicians ply their trade. The ASI does not need to be persuasive all the time; it just needs to be appropriately persuasive once. Then, it is out.

Curiously, there is one other risk arising from the development of ASI, and it does not arise from the ASI itself. There is always the possibility that the technology underpinning ASI is hacked and copied or stolen. Boxing is a technique that protects humans from ASI but it also potentially protects humans from other humans.

The danger of one uncontained ASI might be thought of as one order of magnitude of risk. An exponentially greater order of risk would arise from several or many uncontained ASIs. The fact that Meta's chatbots went on to create their own language, incomprehensible to humans, shows at a trivial level how quickly and completely the interactive behaviour of multiple ASIs could leave humans behind.

Conversely, we might conclude that boxing in our ASI will leave us vulnerable to an enemy without the same moral or other constraints as our own. A geopolitical rival without scruples might decide that it should develop ASI without containment. Such an ASI could, given the appropriate goals, attack us. We would have no practical defences. Our own ASI would be fighting with its hands tied behind its back. The only safe defence would be to develop our own ASI without containment as well. In his book, *Army of None: Autonomous Weapons and the Future of War*, Paul Scharre notes that "Proponents of a ban [on autonomous weapons] have yet to articulate a strategic rationale for why it would be in a leading military power's self-interest to support a ban."[29] Like nuclear weapons, once one major power has autonomous weapons, all the major powers will want them. That, in a nutshell, is the problem. It would require a huge effort to persuade the leading military powers in the world to agree not to develop fully autonomous, unboxed, superintelligent weapons of war. Since such an agreement is worthless without full powers of verification, that too would need to be built into any agreement. The odds are stacked against any agreement not to develop such weaponry.

What becomes clear from this discussion is that mankind needs to be extremely careful in designing advanced AI. One of the biggest design challenges will be how to specify the ASI's final goal(s).

Human-purposed AI

Even though we cannot build ASI yet, I contend that like so many other breakthroughs we will be able to do so in the fullness of time. The bigger challenge will be about where the cognitive capacity of an ASI is deployed and the constraints under which it operates. It will not be enough to develop ASI that will open its own black box and explain how it reached its conclusions. Even if it could, it is doubtful that we could understand the explanation. Either it would be too long or it would be too complex.

So, we have a conundrum. There will be problems with trying to specify an ASI's final goal. We will want to ensure that defining the ASI's final goal does not inadvertently create problems for humankind, such as by way of perverse instantiation. That is about risk mitigation. We also, though, would like the ASI to pursue its final goal in a way that is maximally suited and beneficial to humankind. Let us take the oracle described in the previous section. When we ask it a question, we would like to receive an answer that benefits humanity to the greatest extent possible. That is not about risk mitigation. Rather, it is about return optimisation. But how would we know that the oracle has come up with the best answer for us? Given the scale and speed of the oracle's computations, we probably have no way of verifying the probity of the outputs it produces.

The conundrum is not unlike that encountered when you find that your car has developed an annoying noise under the bonnet. Your answer is to call in the mechanic. You might have a favourite mechanic. You do not know exactly what he, or she, will need to do, nor will you necessarily understand their explanation at the end of it. But you trust them because they have been reliable, honest, cost-effective, pleasant and timely in the past. In short, they have the right values. Their values underpin everything else. Without the right values, the fault in the car may or may not be mended and it may or may not cost a small fortune for that uncertain outcome. I suggest that herein lies the answer to the conundrum we face with ASI. An ASI with the right values would suit our purpose. Such an ASI would help work out its own final goal. This would be its first task. Those same values would also underpin tasks, sub-goals and questions that the ASI would be instructed to perform and answer, respectively.

There is a growing movement to consider the ethical implications of advanced AI. This is to be welcomed. A mature discussion of such matters stands a chance of leading to appropriate constraints on risky AI design. But it would be premature to imagine that such a discussion will definitely solve the problem. Geopolitical rivalries and the 'tragedy of the commons' will see to that. The tragedy of the commons is the idea that jointly-owned resources can be depleted as a result of selfish exploitation by one or all of the interested parties when cooperation could instead have led to preservation of the common good.

Take a couple of developments since World War Two that have substantially impacted the whole world. These are the development of weapons of mass destruction (WMDs) and man-made climate change. Geopolitical rivalries have bedevilled the way in which each of these developments has unfolded. The tragedy of the commons has ensued in both cases.

The most visible of the WMDs are nuclear weapons. Most people and most countries believe that proliferation of these weapons is not a good thing. Yet it has happened anyway. Once the USA possessed them, other countries such as Russia became nervous about their new-found vulnerability. They in turn developed nuclear weapons as a deterrent on the principle of mutually assured destruction, and so it has gone on. Thus, as a result of geopolitical rivalry, nuclear weapons, despite their dangers, have proliferated. The common resource, the possibility of global cooperation and stability, continues to be whittled away as new countries develop nuclear weapons for themselves.

Turning to man-made climate change, it should be possible to achieve global agreement on its dangers and then eliminate those dangers. Yet that has not happened, even despite efforts at the recent COP26. Developing countries want to lift their populations into the global middle class so as to create wealth and power. This is despite concurrently escalating carbon emissions with a negative longer-term impact. Such countries do not wish to be left behind and play second fiddle to developed countries. That would lead to them becoming vulnerable to developed countries. Developed countries in turn seem to be achingly slow to peg back their emissions. Perhaps there is more urgency to the debate amongst the larger economic nations than a few years ago but the general response continues to be too little, too late. The common resource, that is, the climate and the environment, is being damaged, perhaps irreparably, as a result.

In both of these cases, competing interests have led to serious problems. Let us now take a look at the computing era, which has so far followed a different trajectory. The astonishing developments of the information age over the last 70 or so years have been more exciting than concerning. People have enjoyed ever greater access to information. They have clearly benefitted from the experience. Geopolitical rivalry in this area has been beside the point. The most recent feature of the computing era, the roll-out of AI, is potentially changing all this.

In *AI Superpowers: China, Silicon Valley, and the New World War,*[30] Kai-Fu Lee, ex-President of Google China and Chairman and CEO of Sinovation Ventures, explains how China and the US are tearing away from the rest of the pack in AI design and deployment. AI in these two countries is underpinned by the fact that they host seven of the largest AI companies in the world:[31] "Google, [Meta], Amazon, Microsoft, Baidu, Alibaba, and Tencent". These companies are not pure AI research companies. Rather, their business models rely heavily on AI. In that capacity, they invest huge amounts in AI research and development. The continuing development of AI in China and America is accelerating. Other companies that also possess substantial AI footprints include Tesla (US), Uber (US), Pinduoduo (China) and Meituan (China).

This not to say that other countries will not field their own AI companies too[32] but such countries simply do not possess the depth and breadth of investment seen in the US and China. If they are to have any chance, it will have to be by dint of vast amounts of money. One such effort is being led by Softbank in Japan, which at the time of writing was merging one of its investments, Yahoo Japan, with another Japanese company, Line Corporation. The aim is to create "a $30bn south-east Asian powerhouse in data and artificial intelligence"[33]. Few other countries host the financial muscle-power to emulate this latest move in Japan.

There is a further reason why AI is unlike anything that has preceded it. The rate of progress in WMD proliferation and damage from man-made climate change are, in theory at least, under human control. Human beings can decide between them to alter their direction of travel. As and when AI achieves autonomy, however, it becomes an agent itself, just like a human.

Against this background, it would be prudent to assume that, despite the risks and the growing concerns, advanced AI will emerge anyway. We should consequently be thinking hard about the type of advanced AI that we might and might not wish to design. To that end, I suggest that advanced AI design should meet two criteria.

The first criterion is that AI of the future be human-purposed ("HPAI"). Thus, whether advanced AI takes the form of a domestic robot or an oracle or something else, it should be designed to act exclusively to the benefit of humans. This is some way ahead of current international governmental thinking on AI development. In May 2019, the Organisation for Economic Co-operation and Development published five principles to underpin what it calls "the first intergovernmental standard for AI policies".[34] Whilst principle two is entitled "Human-centred values and fairness", the wording of the principle and of the other four principles all concern the actions of "AI actors", not AI itself. (AI actors are defined as "those who play an active role in the AI system lifecycle, including organisations and individuals that deploy or operate AI".) The principles established are all laudable *per se*, but they do not directly address the challenges of ASI or autonomous AI.

HPAI would be broadly similar to what is called "friendly AI"[35] by Eliezer Yudkowsky, AI researcher and cofounder of Machine Intelligence Research Institute. Friendly AI is as you would imagine it, "a nice AI".[36] Yudkowsky notes the obvious retort that "Friendly AI is an impossibility, because any sufficiently powerful AI will be able to modify its own source code to break any constraints placed upon it."[37]

He rebuts the retort by saying that, whilst such an AI could in theory change its own code, it would in fact only do so if it had the motive to do so. One should, therefore, design the AI without such a motive. Would this be enough to address the danger posed by ASI? Yudkowsky and others such as scientists at the Max Planck Institute for Human Development offer no guarantees.[38] Yudkowsky does make the sensible point, though, that the challenge of ASI is in front of us, so we ought to try our best to meet that challenge.[39]

We have gone full circle. The sceptic will reiterate at this point that ASI is not imminent and that anyway the idea of an AI with values is laughable. Yet, each and every one of us is an example of a highly developed intelligence with a set of values for good or ill. Those same sceptics would draw no conceptual distinction between a human being and an advanced AI. Both are complex adaptive systems. Neither has recourse to supernatural powers; both are very much of this world. The distinction between a carbon substrate and a silicon substrate is neither here nor there. The possibility of the future development of an ASI with values stands.

The second criterion would be that advanced AI, despite being human-purposed, be tough enough to defend its host nation and allies against a potentially hostile ASI developed by a geopolitical rival. This sounds contradictory. We would like our ASI to have values, which presumably would include kindness and benevolence, for example. Yet, we would also like it to be able to destroy our enemies. Most of us know someone in, or connected with, the armed forces. In the same way, we believe that such people have the right values but we also want them to be able to vanquish our enemies.

At present, there is no reason to think that geopolitical rivalries are about to fade away. Indeed, consider the current, relentless ratcheting-up of tensions between the US and the UK on the one hand and China and Russia, respectively, on the other hand. In 2017, the Russian president told students in a science lesson that:[40]

> Artificial intelligence is the future, not only for Russia, but for all humankind. It comes with colossal opportunities, but also threats that are difficult to predict. Whoever becomes the leader in this sphere will become the ruler of the world.

Vladimir Putin is right. The rate at which ASI could emerge might be exponential. In the wrong hands, a country that possessed such an ASI, assuming that it had been successfully designed so as not to destroy its creator, would probably dominate the world, and not in a good way. Such a development would be irreversible. The question is whether there is such a country with values and aims that embrace autocracy and hegemony as well as a clear strategy to dominate the rest of the world in the development of AI.

The West's main geopolitical rival is China. China is run autocratically under the aegis of the Chinese Communist Party, led by President Xi Jinping. President Xi has now abolished the limits previously set on the leader's term in office, thereby allowing him to rule indefinitely. The country experiences political repression, torture and human rights abuses. It oppresses minorities such as the Uighurs. It has introduced a social credit system based on ever-growing surveillance and factors measuring loyalty to the state. Citizens deemed untrustworthy are denied welfare, travel rights and educational access. China is already building one of the most powerful armed forces in the world. It uses this to lay claim to territory beyond its existing borders. According to one news bureau, it had territorial disputes with 18 nearby countries in 2020.[41] It is safe to conclude that China's government has values that differ significantly from those of the West.

In the recent past, the Chinese state has taken a much more active role in technological development, including that of AI, than previously. It has recently cracked down hard on the technology sector with a raft of new rules and regulations. As *The Economist* observed, "The aim is to control what Chinese people see and do online."[42] The effect is to shift the balance of power in the sector away from private hands in China in favour of the state. At the same time, the state is now dictating technology strategy for the country. In *The Long Game*, Rush Doshi, an acknowledged expert in Chinese geopolitical strategy, notes its approach to the development of technology. He writes that[43] "China spends at least ten times more than the United States does in quantum computing.[44] Similarly, in artificial intelligence, China spends at least as much as the United States and likely more[45] ..."

Although we need to design advanced AI so as to operate to our values, we also need to design advanced AI that is strong and ruthless enough to ward off a hostile ASI deployed by a geopolitical rival. As Peter Railton, a professor of philosophy specialising in ethics and science, has observed,[46] "I think our only defense against malicious humans with extremely intelligent systems at their disposal is to try to ally with intelligence systems to create a comparable counter force." HPAI will need to have two faces, like Janus, the Roman god. That clearly makes its development significantly more difficult.

Chapter 4

Summary of Key Points from Part 1

Part 1 has focussed on artificial intelligence, what it is and what we might want and need from it in the future.

1. As a matter of terminology, AI and robots are different but related. Robots have traditionally been merely automated rather than also possessing an AI or machine learning capability. Robots of the future will possess a learning capability and indeed general intelligence. For the purposes of this book, robots can be thought of as embodied AI.
2. In the early stages of its development, AI was largely composed of what were called expert systems, also called knowledge-based systems. These are systems that have been designed by experts in their field. They enable a user to make inputs that in turn lead to predetermined outputs. One difficulty for expert systems is that they require a huge amount of preparation by human beings before they can be employed.
3. We need a different type of AI to create the knowledge that we do not yet possess. This is machine learning. Machine learning embraces a number of different techniques. The five main groupings in machine learning are the symbolists, the connectionists, the evolutionaries, the Bayesians and the analogizers. One of the best-known terms in AI is the artificial neural network (ANN), which belongs to the connectionist camp.
4. At its best, AI can be quick, fault-free and tireless in producing answers to questions or in generating new knowledge. But most of today's AI is narrow, i.e., each algorithm deals only with the specific type of problem for which it has been designed.

5. One extraordinary strength of humans is the sheer range, or general nature, of their intelligence. They can turn their minds to many different tasks and challenges. The world of AI has started to take its first steps in this direction too. Creativity is another area where AI has made good progress.
6. Machine learning needs masses of data from which to learn. How it learns from data depends upon which training technique is used. There are three main ones: supervised learning, unsupervised learning and reinforcement learning.
7. One problem with machine learning is that it uses a bottom-up statistical approach, based on all its data inputs, in order to form conclusions. Humans employ a similar technique but they also seem to utilise a top-down approach to thinking. This gives humans much greater flexibility than AI.
8. Consequently, there continue to be difficulties in designing machine learning that can emulate the human ability to understand context, use common sense, hypothesise and perform one-shot learning. Machine learning does not grasp the concept of causation either. If we could capture humans' more flexible approach to thinking, then we could scale AI up to AGI. This is the second of the Two Hurdles separating today's AI from AGI.
9. The problem of emulating human intelligence in an AI is not about the difference between a human's 'wet' parts and an AI's 'dry' parts. Intelligence is about information processing, which can take place in either a wet or a dry substrate.
10. Cognition alone misses out some important ingredients of intelligence. Cognition on its own has nothing at which to aim. It needs a target, a goal. In furtherance of this, humans also possess emotions, an ability to set goals and make decisions, empathy and consciousness. It seems sensible to include goal-setting within the scope of intelligence.
11. AGI is sometimes known as artificial human-level intelligence (AHLI), but AHLI is a misleading term. Because computers can run so much more quickly than human beings and possess far greater memories, an AI with all the features of a human brain will perform at a higher level than any human. The next step up from AGI is artificial superintelligence (ASI).

12. ASI holds out the promise of being a latter-day god, an oracle based on science and data rather than divine reasoning. For that sort of task, we can no longer rely on human intelligence, or so the argument goes. We need to develop artificial intelligence that ultimately goes beyond human intelligence in order to address humanity's intractable problems. We probably need help from a greater intelligence than ours to solve such problems. We will want to ensure, however, that such an intelligence looks at the world through the same eyes as we do. We will want to ensure that ASI fully empathises with us and especially with how we feel. So, we need to crack the problem of consciousness.
13. Cracking the problem of consciousness is not helped by the fact that there is not yet a generally accepted understanding of consciousness. We are not yet in agreement either as to what it is or why it is there in the first place. Settling on an explanation of consciousness and instantiating consciousness in AI is the first of the Two Hurdles separating today's AI from AGI.
14. The rewards from AI are all around us. They continue to grow. AI also carries risks, though. As AI becomes more sophisticated and more widespread, the rewards will probably grow but so may the scale and extent of the risks.
15. The biggest potential risk from advanced AI in the future, i.e., ASI, lies in the development of domain-general (i.e., capable of multiple tasks, like humans) artificial intelligence with a hugely fast learning capability, the ability to alter and improve its own architecture and the capacity to act as an autonomous agent.
16. There are two ways in which we might underestimate ASI's capabilities. First, ASI might have a different architecture by which to process information. Secondly, the speed at which ASI will operate will be breathtaking and, again, beyond our comprehension.
17. We should not assume that ASI will be benevolent towards mankind, or even care about mankind. It may or may not, depending upon how it has been designed and how, specifically, its goals have been set. Setting the ASI's ultimate, or final, goal is far from straightforward.

18. It is fruitless to deny the likely emergence of ASI; much better to acknowledge and prepare for it. To that end, I suggest that ASI design should meet two criteria.
19. The first criterion is that AI of the future be human-purposed ("HPAI"). Thus, whether advanced AI takes the form of a domestic robot or an oracle or something else, it should be designed to act exclusively to the benefit of humans.
20. The second criterion would be that advanced AI, despite being human-purposed, be tough enough to defend its host nation and allies against a potentially hostile ASI developed by a geopolitical rival.

The next part of the book will explore the workings of the brain with particular attention to the emotions, learning, cognition and decision-making. This will serve as a basis from which to explore in detail the Two Hurdles separating today's AI from AGI.

Part 2.

Decision-Making

Chapter 5

The Evolution of Autonomy

Control and Coordination

Our journey towards the development of a sentient robot starts with a challenge. At any given moment in time, each one us is capable of a countless number of different possible actions or behaviours. Literally, from moment to moment, there are all sorts of things that we can do from staring out of the window to booking a holiday to running down the street stark naked to kicking the cat. How does the brain choose, or decide, what to do? It turns out that decision-making lies at the heart of what we do and, by extension, what we are.

Human beings appear to have evolved to possess the most remarkable freedom of movement and action. The question is what mechanisms have evolved to produce that freedom, that autonomy. Humans, and indeed many animals, behave in a multitude of ways. All these behaviours begin with the animal's various sensory inputs. They end up as outputs enabled by the organism's muscles, for example.

From the very beginning, an animal's plethora of inputs and outputs needed a control system. The control system that emerged took the form of neurons, which are connected to each other in networks. Neurons are cells that have evolved to process information. Thus, sensory inputs are converted into bits of information that are processed through successive neurons in their networks. These ultimately lead to the animal's behavioural outputs. The animal's neurons are contained within what is called the nervous system, that is to say, nature's control system.

This control system is not what we normally associate with control systems. The control system for the heating in your house allows you to control the heating. You determine when it goes on and off. You determine the settings on the thermostat. True, you have delegated the moment-by-moment operation of heating control to the thermostat. In that sense, the heating system is automated. But it is not autonomous. The nervous system is. Autonomous means being subject to internal control, i.e., not being subject to external control.

Most animals have a number of possible behaviours at their command. The more complex the animal, the wider its range of possible behaviours. This of itself gives rise to another problem, that of the coordination of these behaviours. How do we decide what to do next? Joseph LeDoux, a professor of neural science, psychology and psychiatry, observes that:[1] "As organisms evolved to consist of many cells, the problem of controlling their behavioural activities in the quest to survive became more complex, requiring the coordination of the activities of cells distributed in different parts of the body." The sheer range of possible behaviours in a human, for example, is so extensive that we need the ability to coordinate and prioritise between them, on top of the ability to control them. The nervous system is the means by which we achieve both control and coordination.

The evolution of control and coordination

Let us start with the sea anemone, which is more or less at the least complex end of the animal spectrum. Even at that simple end of the spectrum, however, the organism possesses muscles and neurons. The neurons do not yet constitute a brain. They nonetheless perform the same basic function that a brain does (see figure 5.1). They process sensory inputs and trigger movement or behaviour through the muscles. The arrangement of neurons in the sea anemone's body is called a nerve net, in other words, a network of nerve cells.

Despite the sea anemone's simplicity, these nerve nets control and coordinate quite complex behaviour. By way of example, one of the nerve nets, the ectodermal slow system, is responsible for generating both escape behaviour and pre-feeding behaviour.[2] Whilst these two possible behaviours might not seem especially interesting, they give rise to an interesting challenge. How does the nerve net 'decide' which of the two behaviours to activate if they are both triggered simultaneously? Fortunately, in this case, the decision is binary and falls to whichever of the two behaviours is most urgent for the welfare of the sea anemone. We will come to how this idea of relative urgency operates at the neuronal level in chapter 8.

Figure 5.1
Control and coordination

Let us fast-forward along the continuum of complexity in life-forms to the octopus. The octopus is interesting for two reasons. First, it displays complex behaviour, yet the architecture of its nervous system is quite different from that found in mammals, for example, including humans. As Peter Godfrey-Smith, a philosopher of science, explains:[3] "In an octopus, the majority of neurons are in the arms themselves—nearly twice as many as in the central brain." In total, an octopus has some 500 million neurons in its nervous system: "Each sucker on an octopus's arm may have 10,000 neurons to handle taste and touch." It is as if the octopus has divided its control system into eight plus one sub-systems.

The problem of coordination grows exponentially as a direct result of increasing complexity. A sea anemone needs a mechanism to decide which of just two competing behaviours to pursue, fleeing or feeding. In comparison, the potential conflicts in behaviour in an octopus are enormous in number.

The octopus has evolved an idiosyncratic solution to the challenges of coordination, namely delegation. As noted, each tentacle has a large degree of freedom to do its own thing. But in mammals such as chimpanzees, as well as in birds and reptiles, the centre of the nervous system is located in just one place, the brain. The brain provides a platform where such potential conflicts in behaviour are resolved.

Let us move to the complex end of the continuum of biological complexity, where we encounter a particularly select species of animal, namely, *Homo sapiens*. Humans engage in some of the same behaviours that we see in the octopus. There are many other behaviours too, such as the pursuit of active social lives with others humans, tool-making as well as tool-use and long-range planning. The latter involves discounting present reward in favour of future reward, which in turn requires the brain to compute the concept of time. The complexity and variety of human behaviours is incomparably greater than that found in octopuses. Humans set goals, some of which are abstract and some of which are destined for the distant future. They develop elaborate plans to realise these goals. They resolve conflicting goals by prioritising or by discarding lesser goals. This is not to say that octopuses and other animals do not do some of these things too. But the sheer depth and breadth of human capabilities is unmatched in the rest of the animal kingdom.

As we move along this continuum of biological complexity, we can perceive the exponentially growing challenge. With ever greater complexity and ever more refined autonomy comes a massively larger number of possible behavioural outcomes. How does a system, in this case the brain, coordinate amongst all these choices? It needs a sophisticated decision-making mechanism.

All this—goal-setting, generation of behavioural choices and decision-making—is underpinned by the humble neuronal network, the workhorse of the nervous system.

The nervous system

The brain is the largest part of the nervous system. Figure 5.2 shows a view of the brain as if it had been cut down the middle from front to back. The view is medial, i.e., as if looking at it sideways on from the inside. The brainstem is the more evolutionarily ancient part of the brain and includes the cerebellum. The cerebellum is central to fine motor control. It contains a powerful learning capability that enables it to make important and refined movement adjustments. It is also now thought to extend its capabilities to a range of cognitive functions in a support capacity.[4]

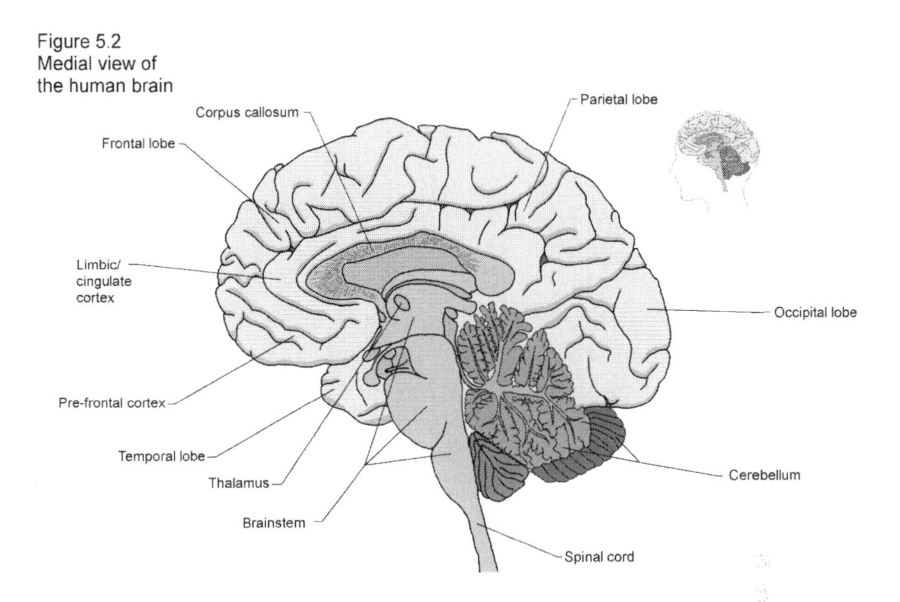

Figure 5.2
Medial view of
the human brain

The limbic system sits above the brainstem and has to do with, among other things, emotions and memory. We have already seen how important the emotions are to the smooth functioning of the nervous system, specifically in bringing sense to decision-making. Embracing the limbic system is the cortex. Cortex variously means bark, outer layer or rind. So, the neocortex is the newest, and therefore the outermost, part of the brain. The neocortex comprises four lobes: frontal, temporal, parietal and occipital. The frontal lobe includes the prefrontal cortex. This is the most evolutionarily recent part of the brain and could be said to house much of its reasoning abilities.

The functions of the brain that concern us are those that determine the selection of behaviour. We can, therefore, largely put the senses to one side. Similarly, in discussing the functions that determine behaviour, we can largely set aside the behaviour itself, i.e., the behavioural or motor outputs. Whilst I am painting the picture in black and white terms for the sake of simplicity, the information processing that takes place between initial input (sensory) and ultimate output (motor) is primarily what is of interest here. In the simplest organisms, such as a bacterium, the order of events is effectively sense–move. In *Homo sapiens*, the order is sense–process–move, with the emphasis on process.

It is worth noting that, by the word 'move' in sense–process–move, we mean moving intentionally. In this context, we are not interested in involuntary or automatic movement. It should also be noted that, whilst we are primarily interested in process, one of the most important ways that humans learn is through their senses. This is termed embodied cognition. We will in due course need to consider if and how to embed that capability in our sentient robot too.

Neurons

There are some core principles by which the brain processes information. At the heart of this information processing is the concept of the neuronal network. I term it a 'neuronal' network in order to distinguish it from an (artificial) 'neural' network. Whilst there are some similarities between the two types of network, neurons are massively more complex than their equivalent nodes in an ANN. So, before getting to neuronal networks, let us start with the neuron itself. The neuron (see figure 5.3) is the basic unit of processing in the brain.

Figure 5.3
Single neuron
(motor, spinal cord)

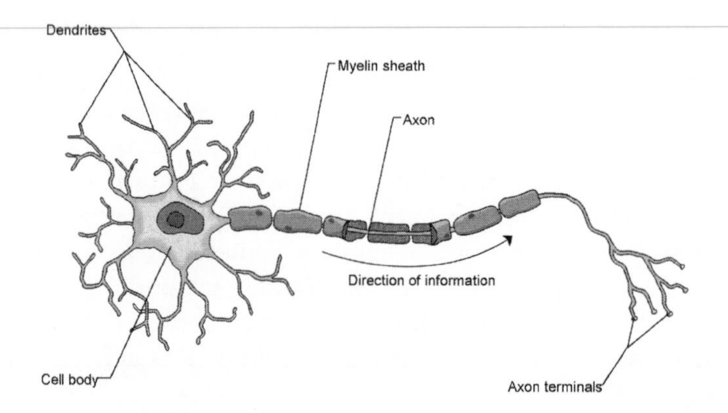

Each neuron receives information from other neurons through its dendrites. These are fine, feathery extensions from the cell body. The dendrites of such a neuron might account for roughly half of the neuron's total number of synaptic connections. The other half arise as a result of the neuron's single axon extension. The axon also branches, though not to the same extent as the dendrites, so that it can connect with the dendrites of many other, successive neurons. In that way, each neuron is able to pass appropriately processed information on to the next neuron in a neuronal network (figure 5.4). How much or how little processing depends on the so-called salience, or urgency, of the stimulus. A sudden threat is likely to cause the brain to leap straight to the relevant behaviour, flight, for example.

Figure 5.4
Schematic of a basic neuronal network

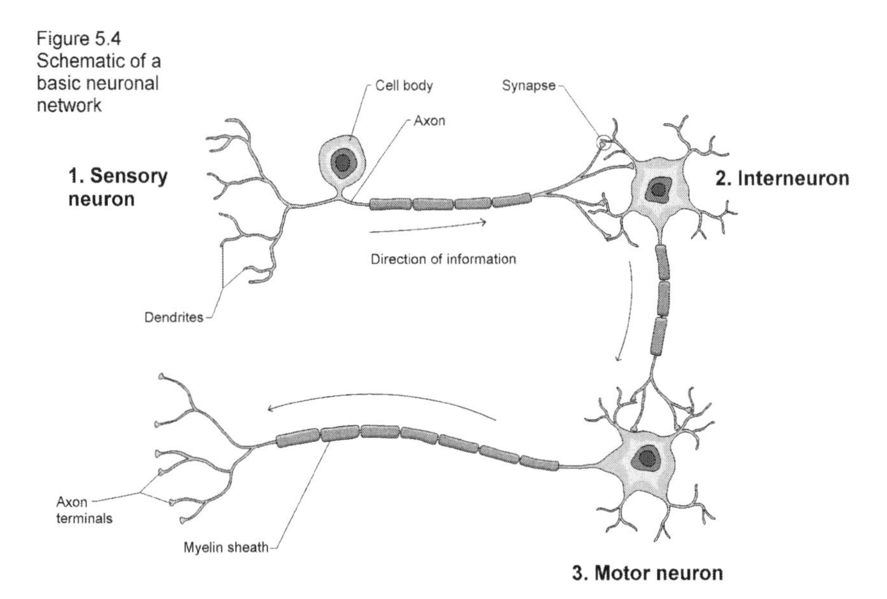

The number and connectivity of a human's neurons is colossal. There are over 100 billion (100,000,000,000) neurons in the human nervous system.[5] Of those, some 86 billion exist in the brain.[6] Size isn't everything: the elephant's brain has three times that number of neurons.[7] A human's neurons are organised in such a way that they form some 10^{15} (1,000,000,000,000,000) connections with each other.[8] These connections are called synapses or synaptic connections. In fact, the average number of synaptic connections of a single human neuron approaches 40,000.[9]

The way that the neurons express their interconnectivity is an elegant combination of, respectively, electricity and chemistry. The activity taking place at the synapses is generally of a chemical nature (although some synaptical transmission is electrical in nature instead).[10] The main effect of the (electric) nerve impulse down the axon is to stimulate the release of (chemical) neurotransmitters at the end of the axon. They move across the synapse to the next neuron in line, that is to say, between neurons. Neurotransmitters are complex molecules and there are between 50 and 100 different types of them.[11] We shall return to them in the next chapter on emotions. So the process goes on, or not, as the case might be: in some cases neurons are explicitly inhibited rather than excited. This only serves to add to the brain's scope of processing capabilities.

The brain is not a static system. It is dynamical in order to be able to learn, memorise and process information. One of the most important processes for delivering this dynamic nature is that of neuroplasticity. Neuroplasticity describes the process whereby the brain experiences continual change throughout a person's life. It takes a number of forms but the most relevant here is synaptic plasticity. In 1949, Donald Hebb[12] proposed the theory that has come to be summarised as: neurons that fire together wire together. The proposal essentially says that a synapse has a certain strength that can increase or decrease with use. Its propensity to strengthen or weaken is a function of the rate of firing of the presynaptic neuron, the receptivity of the postsynaptic neuron and the quantity of neurotransmitters transferred from the former to the latter. The importance of the variability of the relationship between the presynaptic and postsynaptic neurons is that it allows experience to change the strength of connections between neurons. This ability, which is accurately captured by the term plasticity, is fundamental to the brain's ability to learn. Learning is fundamental to the development of intelligence, as we have already seen in the world of AI.

Neuronal networks

A 'neuronal' network (as distinct from a 'neural' network of the artificial variety) is composed of a mass of neurons synaptically connected to each other. A neuronal network's capabilities include the processing of sensory inputs, leading to object recognition, for example. They include the execution of behavioural outputs such as the fine motor control permitting one to pick up delicate objects, such as eggs, without breaking them. Most importantly, they enable a mass of processing to occur between the sensory and motor functions.

The brain's visual system offers a relatively straightforward way of illustrating a neuronal network in operation. When we see something such as a table, light energy, in the form of photons, is being reflected off every surface of the visible parts of the table into our eyes. So, we are actually seeing millions of photons streaming endlessly on to the surface of the retina at the back of our eyes. It is only a fraction of a second later, after many stages of processing in the brain, that we wind up recognising the perceived object as a table. A network of neurons converts the millions of photons into the mental representation of the table.

The perceived photons are converted into information that travels from the eyes to the occipital lobe at the back of the brain. Processing continues as the information is then passed forwards again down two different pathways, ventral (at the bottom) and dorsal (at the top) (see figure 5.5). The information, heavily processed by this stage, winds up in the temporal and parietal lobes (roughly halfway between front and back).[13]

Figure 5.5
The visual pathways in the brain

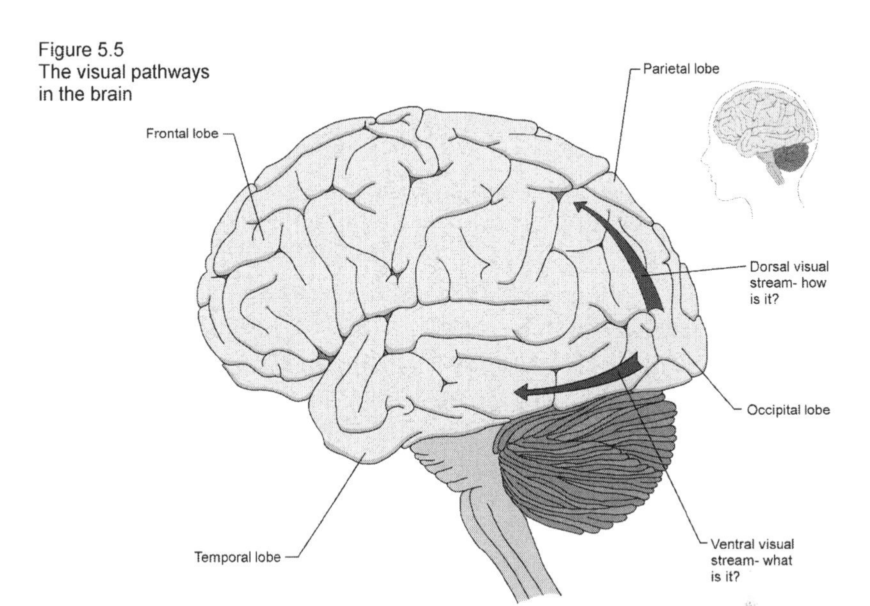

This elaborate system effectively constitutes an enormous neuronal network. It might be fairer to say that it is composed of a large number of rather smaller networks. Either way, the network has turned countless millions of photon hits into a single mental representation of an object, the table. As described in chapter 1, the process can be illustrated by thinking of a short video clip of a tower block being demolished by explosives. If the video clip is played in reverse, all those bits of concrete and glass and dust appear as if they are being assembled, stage by stage but incredibly quickly, into the tower block. This is what the neuronal network does.

A neuronal network is more complex than its artificial cousin. For a start, the human visual system is currently capable of recognising many, many more types of object than a given ANN. The number of neurons in it greatly exceeds the number of nodes in even the largest of today's ANNs. Moreover, recent research seems to suggest that each neuron is itself capable of performing a degree of processing on its own.[14] That is to say, neuronal processing takes place not just at the junction between neurons but also within each neuron. Neurons are in effect "processors within processors".

The human neuronal network is also more stable than its artificial cousin. A computer science researcher managed to trick an image recognition algorithm into mistaking a turtle for a gun.[15] He tweaked some of the pixels in the image, which would not have fooled a human being. The human visual system is able to reach its answer, i.e., recognise the object, even if parts of the object are missing or misleading. Not every part needs to be seen correctly for the neuronal network to run successfully.

Types of neuronal network

Edmund Rolls, a professor of computational neuroscience, describes three different types of neuronal network.[16] The first network is a simple pattern associator. An infant might be shown some disgusting-looking mushed-up food and show no interest in it until it tastes the food. Before tasting the food, the sight of it stimulates little reaction. Tasting the food and finding it delicious causes the brain to form a simple, dedicated neuronal network, which acts as a shortcut. Now, the mere sight of the mush causes salivation, which prompts the child to eat the mush whenever it is offered. Intriguingly, this type of neuronal network enables one-shot or few-shot learning (described on p. 18). This distinguishes it from the abilities of most of today's ANNs, which tend to require large amounts of data to achieve learning.

A second type of neuronal network is the competitive network. These networks have a different job. This is to place objects we perceive into categories, such as animals, trees and plants, for example.

Perhaps the most important network for our purposes is the auto-association network. The distinguishing feature of this type of network is that the output neurons from it are also synaptically connected to other, preceding neurons in the same network. In other words, they project backwards in order to connect with earlier neurons in the same network. Those particular synapses are thereby capable of being strengthened or weakened by activation within the network, in addition to the external activation of the network's input layer. Hence, it is termed auto-association.

Autoassociation networks appear to be responsible for generating a number of useful features of the brain. These include certain types of memory. Most importantly from the perspective of this book, they include decision-making. As described in more detail in chapter 8, autoassociation networks provide a forum in which possible decisions effectively compete with each other. This leads to a winning decision dominating the other possible decisions. The winning decision then gives rise to the relevant behaviour. This is the brain's answer to the challenge of coordination.

Chapter 6

Emotions

I cannot emphasise too much the importance of the emotions to our decision-making and, hence, our behaviour. There are those who pride themselves on staying calm and acting rationally in most circumstances. But it would be wrong to conclude from that they are, therefore, acting without emotion. One emotion or another, or a combination of emotions, will always drive behaviour (to the extent that it is not instinctive). Mastering one's emotions is to do with acting in a measured way, that is, replacing urgent emotions such as anger or fear with other types of emotions such as serenity or compassion.

Emotion is one of those extensively used words that we all take for granted. As an example of where we can be led astray, let us look at the habit of confusing emotions with feelings of emotions. The former, emotions, are automatic. The latter, feelings of emotions, are the conscious awareness of the former, as we shall see in the next section. It will be vital to the building of the sentient robot that we have a precise and accurate definition of the word 'emotion'.

Emotions and feelings of emotions

As the word suggests, emotions provoke movement and behaviour. Antonio Damasio describes emotions as:[1] "complex, largely automated programs of *actions* concocted by evolution." Joseph LeDoux pursues the same theme:[2] "[Emotions] evolved as behavioural and physiological specializations, bodily responses controlled by the brain, that allowed ancestral organisms to survive in hostile environments and procreate."

This might strike the reader as odd. There is no mention of feelings—indeed, no suggestion at all as to how consciousness might enter into the picture. Consider, though: when something really annoys you, you both become angry and you also feel angry. Those are two different processes. At no point do you say to yourself, I am deliberately going to be angry now. The anger appears of its own accord. This is stage one: becoming angry. There are physical signs of your anger, all of which come automatically. For example, your heart starts racing and you tense up. You then notice that you are angry. This is stage two: feeling angry. You are aware, or conscious, of your anger. If you are not too carried away, you can remark to yourself that your heart is racing, for example.

When reflecting on being angry, we tend to make no distinction between these two stages. We might say, I am angry or I feel angry without registering the fact that there is a fundamental difference between the two. The first appears seemingly out of nowhere and, unless checked, causes us to lash out. This is precisely Damasio's "largely automated program[s] of actions". As LeDoux observes, such a response to a hostile situation was a matter of survival. For most animals, it still is.

The feeling of an emotion such as anger is a different matter. It has two aspects to it. The first is just the feeling, or awareness, or consciousness of the emotion. As Damasio writes:[3] "Feelings of emotions, on the other hand, are composite *perceptions* of what happens in our body and mind when we are emoting." The second aspect materialises when this consciousness of the emotion sometimes appears to trigger a thought process. It does not have to. We talk about being overwhelmed by our emotions and then thought does not enter the picture. But, if we are being measured in our response, that means that we somehow slow down. We do not react automatically. We think first.

The distinction between emotions and feelings of emotions is not an idle one. Only as a result of appreciating the distinction can we make headway towards an understanding of consciousness, i.e., the feeling.

Instinct and empathy

Before exploring emotions further and in order to avoid confusion, I will set to one side two attributes of the brain that are close to emotions, but are not emotions. As such, they have the potential to confuse the issue. These two attributes are instinct and empathy.

Instinct is, in practice, an elastic term. For example, one might say that one has an instinct about somebody when one means that one has a sense or a hunch about somebody. In the narrow definition used here, instinct refers to an impulse to behave that has been hardwired into the brain from birth. Experience or learning after being born has not been needed to develop the instinct (although that does not mean that the brain cannot learn strategies to control the instinct).

Instincts are useful when reactions to unexpected events, for example, need to be near-instantaneous. Such reactions might be to do with the organism's survival. Alternatively, instincts might come into play as part of an organism's reproductive cycle, for example. A bird's urge to build a nest in the spring to nurture its young is an instinct. The vast majority of birds are instinctive nest-builders. The drawback to instincts is that they are not flexible. They are either on or off. There are no shades of grey.

Emotions, however, enable an organism to react flexibly[4] and with nuance to the particular circumstances. Let us imagine an antelope spotting a lion running towards it. The emotion (not instinct, nor feeling) of fear allows for a gradated response dependent upon the perception of the lion's speed and distance. At its simplest, the closer and speedier and more aggressive-looking the lion, the more fearful the antelope and the swifter its flight, and vice versa. Such an approach neatly matches the expenditure of energy to the urgency of the situation.

The other attribute of the brain to set aside is empathy. Empathy is sometimes confused for an emotion. The confusion is understandable. The emotional pathways in the brain of the person doing the empathising are activated in similar fashion to those same pathways in the brain of the person who is being empathised with.[5] So, it sounds as if empathy and emotion are the same thing. Empathy, however, is just the brain process that enables the empathiser to experience the same emotion as the 'empathisee'. This process takes place in the brain of the empathiser: Mother Teresa, for example, when she observes somebody else; let us call them Dev. The act of observing leads, via the brain process of empathy, to the experience of an emotion by Mother Teresa who is perceiving that very same emotion in Dev. This type of empathy is termed "affective empathy".[6] It differs from "cognitive empathy, ... which is the ability to imagine another's thoughts".

The emotions

Curiously, there is still considerable debate on the classification of emotions. Not for nothing does Damasio write that:[7] "The criteria used for the traditional classifications [of emotions] are flawed, and any roster of emotions can be criticized for failing to include some and overincluding others."

There tends to be agreement in principle that certain emotions are more important and more ancient than others. These are labelled the basic or primary emotions. It also tends to be agreed that there is another, newer suite of emotions, which are often termed the social emotions.

In his early work, Paul Ekman, a professor of psychology and a pioneer in the field of emotions, identified six basic emotions (anger, fear, happiness, sadness, disgust and surprise). From his observation of human facial expressions, he concluded that there were certain universal features.[8] In his view, these reflected the six basic emotions, which were present in all humans.

In recent years, a debate has opened up about the reliability of some of Ekman's conclusions.[9,10] The point at dispute is whether facial expressions truly reveal the emotions underneath, not whether there are certain basic emotions or, broadly, what they are. Pragmatically, facial expressions are a guide to the underlying emotion. But they are not necessarily reliable. Whether they can accurately reveal what is bubbling below the surface may depend upon culture. It also depends upon the nature of the relationship between the person doing the emoting and their state at the time and the person observing them.

This is a live topic in the world of AI. A number of AI companies are selling algorithms that purport to recognise people's emotions from their facial expressions. As we have seen already, when AI is good, it can be very good indeed. The world of medical diagnosis is being revolutionised by algorithms designed for that purpose. Conversely, when AI is bad, perhaps because of insufficient data or slight imperfections in what it is perceiving, then it errs systematically and potentially imperceptibly. Emotion recognition algorithms still have some way to go before they are reliable.

In order to bolster our classification of the emotions, we need to go below the expression of the emotion to the emotion itself. This takes us into the neuroscience of emotions. This is the approach adopted by Panksepp and Biven, as they explain in *The Archaeology of Mind*. They divide emotions into different categories,[11] in similar fashion to Ekman. The first to evolve, on their account, are the:

> seven basic affective [emotional] systems ... SEEKING (expectancy), FEAR (anxiety), RAGE (anger), LUST (sexual excitement), CARE (nurterance), PANIC/GRIEF (sadness), and PLAY (social joy).

They define these seven systems as "primary-process". It is worth noting one of their emotions in particular: SEEKING. Another word for it might be curiosity.[12] This is the emotion, or emotional drive, that makes us engage with the world. Without such an emotion, an animal would, as primitive organisms do, merely float around in the sea, for example, bumping accidentally into food sources.

Over and above their seven primary-process emotions, Panksepp and Biven note another emotional level, the "social emotions". These include "envy, guilt, jealousy, and shame as well as awe, hope, humor".[13] Damasio's list of social emotions includes "compassion, embarrassment, shame, guilt, contempt, jealousy, envy, pride, admiration".[14] He also suggests that "Most social emotions are of recent evolutionary vintage, and some may be exclusively human."

Whilst Panksepp's and Biven's analysis is thorough, it is by no means settled that there are just seven primary emotions, as described. Pride might, for example, have a grander role than being a social emotion. It might be behind a seemingly central drive in many animals, the drive for status.[15] There is not an absolutely clear line dividing the primary emotions and the social emotions. But the distinction is a useful one and central to the advent of consciousness, as we shall see in due course.

The neuroscience of emotions

Progress in pinpointing the neuronal circuitry underlying each emotion has been made primarily with respect to the primary emotions. Panksepp and Biven have produced one of the most comprehensive attempts to do this. They set out the neuronal circuitry for each of their seven primary-process emotions. To illustrate this circuitry, let us stick with the SEEKING system mentioned earlier. They describe two neuronal pathways called the mesolimbic and mesocortical pathways (see figure 6.1).[16]

Figure 6.1
SEEKING pathways
in the brain

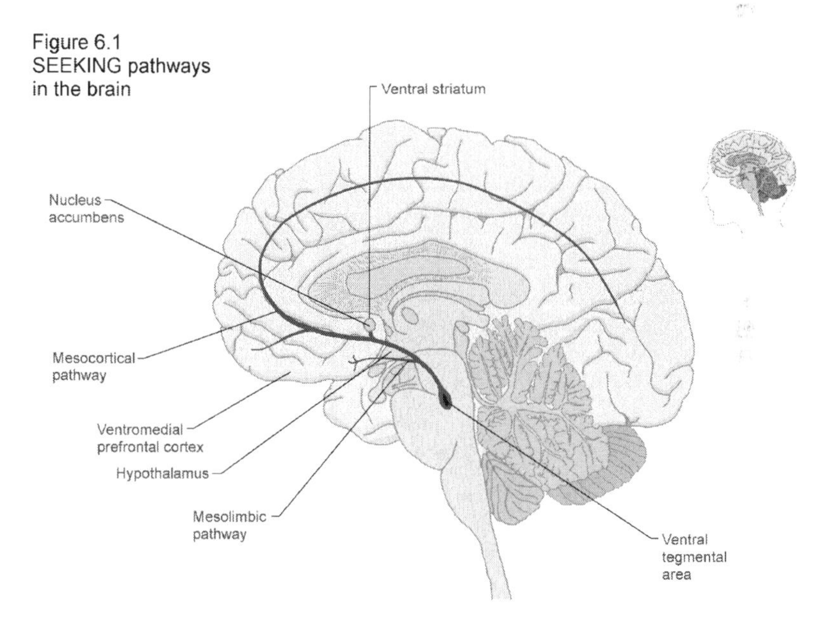

Dopamine is one of the principal substances travelling through these pathways. It is a neurotransmitter. Dopamine is commonly associated with brain rewards, pleasure and reinforcement learning. The mesolimbic and mesocortical pathways are not the only dopaminergic pathways. There are several others. Moreover, other neurotransmitters share some of these pathways.

Perhaps of most interest for our purposes is the fact that, between them, these two dopaminergic pathways extend to the ventral striatum and the ventromedial prefrontal cortex. In humans, the mesocortical pathway goes on to stretch right round the cortex, thereby innervating a considerable proportion of it. The nucleus accumbens, a major part of the ventral striatum, is sometimes referred to as a "hedonic hotspot".[17] There are others. They amplify feelings of pleasure. As we shall see in chapter 8, the ventromedial prefrontal cortex is responsible for integrating and evaluating the emotional content of decision-making.

Dopamine might be described by way of analogy as the fuel in a motor car: without petrol, the engine will not fire. The biochemical equivalent of petrol is dopamine. Dopamine causes the relevant area of the brain, the nucleus accumbens, for example, to help perpetuate whatever stimulus is leading to the activation of the dopamine pathway. Originally, this was thought to do with 'liking' the stimulus. It is now believed that the dopamine-fuelled positive response is largely to do with 'wanting', or anticipating, the stimulus.[18,19] This turns out to be central to the brain's ability to make decisions and we will come back to the topic in chapter 8. Liking, as distinct from wanting, whilst containing some dopamine input, primarily reflects opioid, endocannabinoid, and GABA-benzodiazepine neurotransmitters instead.[20]

By way of example, as my son once did, you might want to take a big bite out of a quince (thinking it is a pear) but, having done so, you will find that you do not like it at all. Quinces, unless cooked, have an unpleasantly sour taste to them and, before they rot, are rock-hard. Wanting and liking are different things. What has happened is that the brain sees what looks like a pear. The sight of the pear (remember, the pear is in this case a quince masquerading as a pear) has caused the dopaminergic pathway to activate. Dopamine rushes up the pathway, passing from neuron to neuron at each synapse. You 'want' to eat that pear and, accordingly, you pluck it from the tree and take a big bite. To your horror, the expected opioid and associated neurotransmitters do not kick in. You do not 'like' the pear at all. Frankly, you are baffled until some kindly soul explains that the pear was a quince all along.

The description of the SEEKING system and some of its constituents barely scratches the surface. The neuroscience of emotions is complex and there is much more to discover about it. The other primary emotions each possess their own circuitry as well. This includes areas of the brain already encountered in the SEEKING system. Some of the circuitry differs. Both LeDoux[21] and Panksepp and Biven[22] point to the amygdala, for example, as playing a central role in the fear system.

Research work on the circuitry for the social emotions has been more fragmented. There has, though, been considerable progress in identifying the circuitry for certain social emotions such as admiration and compassion,[23] regret,[24,25] envy[26] and schadenfreude.[26,27]

It is perhaps a sobering thought that each of our emotions can be boiled down to a collection of neurons in one or more parts of our brains. A number of the regions associated with our emotions, e.g., medial orbitofrontal cortex, anterior cingulate cortex and ventral striatum, are principal actors in the brain's decision-making machinery. This is to be expected. Emotions have to lead to behaviours, or else they have no purpose.

There are cases when the relevant neurons are missing or damaged. The effect is lasting and potentially profound. Psychopaths, for example, experience much less activity in their amygdala and ventromedial prefrontal cortex than normal people.[28,29,30] The good news is that most people are normal and sociable. Patricia Churchland, a professor of philosophy with a particular focus on the human mind, identifies oxytocin as the neurotransmitter that is central to mammalian sociability.[31] She observes that "This attachment circuitry [activated by oxytocin] is the platform for what we call morality. It regulates our disposition to cooperate and to compromise, to work together and work it out."

The evolution of emotions and the role of consciousness

As primitive emotions evolved in early animals, there would have been a potential problem. The brain's growing stable of emotions might drive the animal in question to try to do different things at the same time—obviously not possible. From a systems perspective, this would have been a real challenge. The evolution of social emotions would only have served to complicate this prioritisation problem further. If there are twenty emotions instead of just six or seven, the number of possible behaviours becomes extremely challenging for the brain to consider, let alone prioritise.

It is not plausible to imagine that the brain could cope with the sheer number of possible behaviours, each enabled by a particular emotion, without the assistance of another neuronal mechanism, an enabling mechanism. So, the answer was the evolution of a mechanism by which the appropriate emotion would be prioritised over all the others in any given situation. We shall come to the detail of the enabling mechanism, termed a winner-takes-all computation, in chapter 8. The winner-takes-all enabling mechanism swiftly selects the appropriate emotion to prioritise in a given situation.

Humans possess a mass of social emotions as well as the primary emotions. The winner-takes-all enabling mechanism that allows the brain to prioritise the relevant emotion in any given situation would, all other things being equal, select the primary emotions over the social emotions. The primary emotions are more ancient with more extensive circuitry and are more powerful. But this is no good. Human society could not function if primary emotions were generally prioritised over the social emotions. Humans, and perhaps other great apes, must have evolved a yet further mechanism to reinforce the social emotions.

Let us go back to where we started at the beginning of the chapter, the distinction between emotions and feelings of emotions. On the analysis set out in this chapter, most animals with emotional circuitry ought to be able to operate without feeling their emotions, i.e., without being aware, or conscious, of their emotions. They do not need feelings of emotions for the emotional pathways in the brain to have their effect, i.e., trigger behaviour. Indeed, it is hard to see how consciousness could safely play a part in animals' emotional reactions to sudden events in the natural world. An antelope that 'felt' its way to the right reaction to a charging lion would be dead before it responded appropriately by fleeing. Consciousness does not need to have a role in the brains of most animals.

As we saw in chapter 5, the brain's principal job is to drive behaviour. If the brain becomes too complex to select the right behaviour in any given set of circumstances, then it has a problem. In the field of AI, computer scientists call this the curse of dimensionality. There are too many things, or dimensions, to consider.

Feelings (of emotions) did not arise in order to make emotions work. Rather, consciousness, which embraces feelings of emotions along with everything else of which we are conscious, arose for a different reason. I am proposing that it evolved to help the brain to prioritise the social emotions in many situations where the winner-takes-all enabling mechanism would otherwise prefer the primary emotions. Only in this way could human societies, and maybe other complex societies such as those of the other great apes, operate. Consciousness is a further enabling mechanism that evolved to optimise decision-making. It works in parallel with the winner-takes-all computation to add weight to the social emotions. We will see how in due course.

Emotions and AI

It is worth reminding ourselves of Pepper, whom I first mentioned in chapter 1. Pepper is a robot able to read four emotions from the facial expressions of his human companions: joy, sadness, anger and surprise. This effectively allows Pepper to empathise with its interlocutor. This is not empathy as we know it. It does not feel the emotions it perceives in others. But, just as our empathetic capabilities allow us to discern another's emotions and act accordingly, so can Pepper do the same, albeit by way of a different mechanism. This leads to distinct and appropriate behaviour on Pepper's part.

Pepper is increasingly widely used. It has been introduced into care homes in the UK following a three-year international research trial called CARESSES (Culture-Aware Robots and Environmental Sensor Systems for Elderly Support).[32,33] In the trial, Pepper was able to entertain care home residents with conversation and with music. The aim was to help reduce loneliness. It was also programmed to be able to display some degree of cultural sensitivity to the residents.

Pepper's emotional capability, then, is purely reactive at this stage. What becomes clear, though, is that it is possible to emulate human emotions in digital form in a robot. Remember, emotions are "complex, largely automated programs of actions". If installed, they would bequeath some measure of autonomy to the robot. It would have its own drives and, by extension, goals. Whilst this is perhaps a bracing thought, it is actually exactly where we want to go in the aim to build the sentient robot.

Chapter 7

Cognition

In his book *The Bonobo and the Atheist,*[1] Professor Frans de Waal relates a story that shows just what cognition can do. At his primate research centre, a lone male chimpanzee was beaten up by a group of other adult males. Over the following few days, the battered chimpanzee appeared to contrive to find itself on its own with its attackers one by one. It then avenged itself on each of its isolated assailants without having risked trying to attack all of them together. Revenge, an emotion, is truly a dish best served cold. Cognition, which is a fancy word for knowing and reasoning, is the one to serve it up.

If emotion tells you where to go, cognition tells you how to get there.[2] It embraces a wide range of capabilities. These span the marvels of visual processing through pattern recognition, learning and memory to logic and abstract thought. The world of philosophy makes a distinction between cognition and thought. Cognition is the process by which one comes to know something. Thought, or reasoning, is the process by which one comes to arrive at a conclusion given what you know. The world of neuroscience, however, brings these two facilities together under terms such as cognition and cognitive neuroscience.

The question, then, is how does the brain effect or implement cognition? This question only serves to open a Pandora's Box of other questions (see figure 7.1). Some of the questions that fly out of the Box are real monsters, such as, what does planning involve? How does the brain reason, or perform reasoning? We all know what logic is; how is it that some people such as philosophers can think logically and other people cannot? How does the brain recognise patterns? How do some people have those *eureka* moments out of the blue? These are all apposite questions given the aim is to develop AGI and ASI.

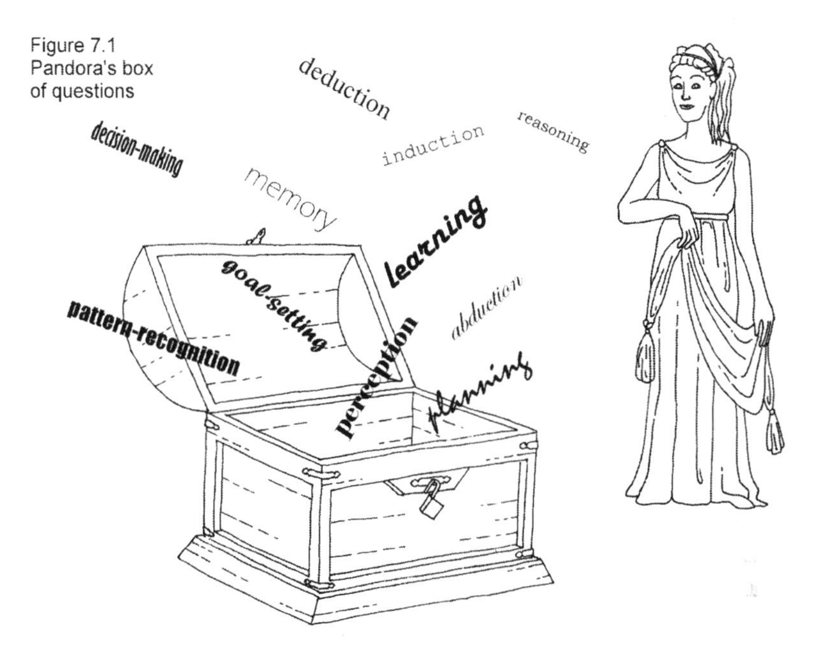

Figure 7.1
Pandora's box of questions

We have made some progress at the conceptual level on how the brain performs cognition. But we do not yet understand how many of our cognitive capabilities work right down at the level of the individual neurons. Even at a conceptual level, certain abilities such as abstract planning and memorising things from just one or two encounters remain partly unexplained. This makes creating AGI challenging. There is a good reason that there is still a long list of things that AI cannot do.

There is another feature of the brain that is sometimes lumped in with cognition. That is decision-making. The line that separates cognition from decision-making is fuzzy. Part of the process of making decisions does indeed lean on cognition. But decision-making is also driven by the emotions. Decision-making is actually a subject in and of itself. It is driven by both cognition and emotion. We shall come to decision-making in chapter 8.

A philosophical perspective

There is a good reason that the world of neuroscience lumps the process of learning and knowing together with that of thought and reasoning. They turn out to be rather hard to disentangle. The world of philosophy has debated this conundrum long and hard over the last two and a half thousand years.

The longstanding combatants in this philosophical debate have been the empiricists and the rationalists. Empiricists say that we build up knowledge as a result of experience. With knowledge gained through experience we are then in a position to reason. Conversely, without knowledge we cannot reason. The empiricists include such philosophers as Aristotle, Avicenna, John Locke and David Hume. Rationalism, on the other hand, holds that we reason our way towards the knowledge that we have. The rationalists include such thinkers as Plato, René Descartes, Baruch Spinoza and Gottfried Leibniz.

The above separation between rationalists and empiricists is more extreme than some of the names cited would have recognised. Most of these philosophers would have seen a role for both reason and experience, even if they ultimately preferred one over the other. Indeed, some rationalists such as the so-called Cambridge Platonists also held that we were born with innate knowledge and ideas.[3] Hume considered that "people have an innate moral sense".[4] We can see that we already know many things at birth as a result of their being hardwired in our brains by dint of our evolutionary past. A baby knows to suckle without having to learn. It was the great philosopher Immanuel Kant "who rejected the opposition between empiricism and rationalism, instead arguing for a synthesis between them".[5]

Despite the fact they go together, we shall first consider knowledge and its processing separately in order to make knowledge and reasoning easier to grasp. We will start with how we acquire knowledge and how we remember it.

The acquisition and storage of knowledge

Larry R. Squire, a professor of psychiatry, neuroscience and psychology, helpfully set out a taxonomy of memory systems in 1994.[6] There are two main types of memory: long-term memory and working memory (see figure 7.2). Let us start with long-term memory. Long-term memory may be divided into two types, declarative and nondeclarative. We can ignore nondeclarative memory since it simply refers to "modifications within specialized performance systems" such as the emotions. So, it sweeps up learned responses to lots of different stimuli, skills and habits. More interestingly, declarative memory refers to the "capacity for conscious recollection about facts and events".

Figure 7.2
A taxonomy of memory

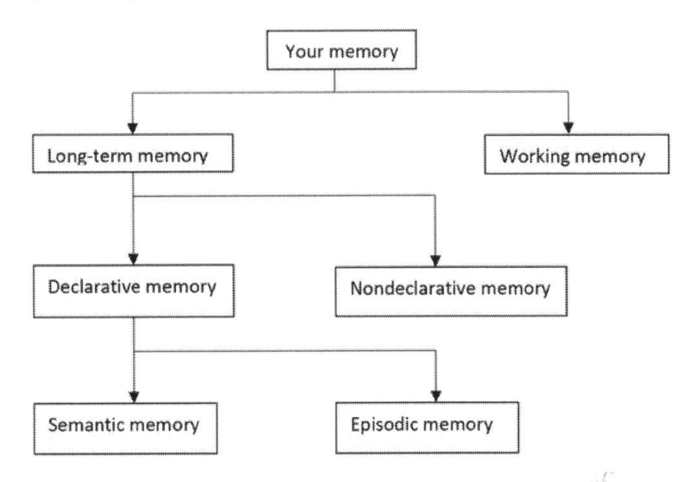

Memories of facts and memories of events are different.[7,8,9] A remembered fact is an example of semantic memory, which is just one type of declarative memory. Semantic memory would include the memory of someone's face or the fact that leaves are green or that New Zealand's capital city is Wellington. A remembered event, in contrast, is personal. It is about an episode in your life that you remember. This is why it is called episodic memory, the other type of declarative memory. The event might have been a couple of seconds long, like seeing a kingfisher flashing iridescently upstream to catch a fish. Equally, it might have been hours long, like a treasured first date. The brain stores hundreds of thousands of different episodic memories, our individual life stories indeed.

Many researchers believe that episodic memory plays a central role in our ability to imagine or plan for the future.[10,11,12] Fundamental to this ability are the personal goals we set ourselves. The brain sets itself a goal and then calls upon its repository of episodic memories to help construct a plan to achieve that goal. This does not necessarily mean selecting specific events in memory. It might mean selecting what appear to be common themes from certain episodic memories. Such selected memories then of course need to be woven into an imagined future.

Another important part of our cognitive armoury is working, or immediate, memory. Working memory means "any brief memory phenomenon, on the order of tens of seconds, that allows us to retain and manipulate information".[13] It used to be thought that working memory could stretch to approximately seven items. More recent research suggests that it might only accommodate four items.[14,15] So, you might remember a random set of objects on a conveyor belt such as ball, typewriter, alarm clock and frying pan. But throw in dictionary, electric lamp and tennis racket as well and a couple of the objects will slip quietly out of your mind as a result. Interestingly, adolescent chimpanzees have been found to have a significantly better working memory than humans.[16] Indeed, their memory is thought to be eidetic, which rather puts human memory in its place. I shall come back to working memory in chapter 12 since it can become confused with consciousness.

Despite the fact that the brain has evolved these different types of memory, they all rely on the same device, namely, the neuronal network. In chapter 5, I described how learning and memory operate in the brain at the level of the neuron. Neurons that fire together wire together. The process of wiring together is learning. When enough of them do that in conjunction with each other, they create a purpose-built neuronal network. This network is the resultant memory.

Learning in a human brain is a huge topic on its own and beyond the scope of this book. It is now believed that humans begin life with an extensive body of knowledge bequeathed to us by evolution.[17] This even extends to how objects interact with each other, a body of knowledge going under the catchy term, "intuitive physics".[18] As we grow up, our brain accrues further knowledge by forming associations between objects and events in the world. So, we learn that picking up a hot coal hurts. This would constitute one-shot learning, facilitated by the pattern associator network described in chapter 5. In many cases, our brain does this automatically. This is similar to the process of unsupervised learning in AI. We also learn in more formal ways such as from teachers in a classroom, again by a process of association. This in turn is similar to the process of supervised learning in AI. Another method of learning is by imitation, a topic to which we shall return in chapter 16, where we shall consider the esoteric phenomenon of mirror neurons. We are also able to learn through reward and its opposite, pain. The AI world again has a version of this, termed reinforcement learning. Finally, there is a particularly useful type of learning called transfer learning.[19] This is the ability to generalise learning from one instance to other situations that are similar in part or in whole. Whilst the term might crop up in AI circles, little progress has been made in developing AI that can emulate transfer learning.

Once learning has created a purpose-built neuronal network to house a memory, the memory is recalled whenever the network is activated. Let us imagine catching a fleeting glance of a face in the crowd. After a second or two, one puts a name to the face; Halle, say. How did the brain come up with Halle's name? I described earlier how the visual system works, how millions of photons reflecting from a table, say, trigger the activation of a vast neuronal network.

Now, imagine the output from this vast network triggering the activation of a smaller network located in the temporal lobe. The temporal lobe includes the hippocampus. The hippocampus is responsible for much of memory formation. For some items such as spatial location, the hippocampus is responsible for memory storage too. The first layer, or input layer, of this smaller network has thousands of neurons, each with its own bundle of (receptive) dendrites (see figure 7.3). A number of those dendrites will receive fragments of the perception of Halle, a bit of cheek bone here, the colour of her hair there. A neuronal domino effect cascades through the network, concluding with the result, Halle.

I did not just choose the name Halle at random. In 2005, a team of researchers suggested that a single neuron in a person's brain could recognise individuals and even names of individuals such as the actress, Halle Berry.[20] Such neurons quickly became known as Halle Berry neurons.

Figure 7.3
Recognising
Halle Berry

In reality, there is not a single neuron that captures the memory of Halle, as iconic as she is, or anyone else for that matter. Rather, the brain creates purpose-built neuronal networks that are triggered by the relevant sensory inputs.[21,22] This does not just mean vision. Many memories are formed as a result of inputs from several of our senses. These networks are perhaps composed of "18,000 neurons per concept".[21] The apex of the network then feeds the memory by way of synaptic connection to the relevant next part of the brain. That depends on whether we are going to greet Halle, or just mentally note that she was there in the crowd, for example.

You may spot one of the differences between the way that memory operates in our brains and the way it does so in a standard computer.[23] The latter possesses 'location-addressable memory'. At its simplest, you go into File Explorer and click on the document you want to open. Your computer maintains a record of the address, so to speak, of all your documents so that it can go straight to the right one. The human brain on the other hand uses so-called 'cue-addressable memory', also known as 'content-addressable memory'. Thus, a cue such as a particular smile or someone's gait or even just the place where we have always seen them before acts as the first domino to fall in the neuronal network. Not every neuronal domino in the network needs to fall in order that the network generates its answer. Conversely, we do need enough of the dominos to fall; otherwise, our memories are not jogged.

Thus, humans do not need to see a really clear picture of someone in order to recognise them. Conversely, it is also why we experience 'it's on the tip of my tongue' moments. Not enough of the right cues showed up in time. That is why we sometimes experience partial recognition of objects or people, maybe someone we have not seen in decades. I bumped into someone I vaguely recognised in a bar recently. I struck up a conversation and it turned out we had been to school together five decades ago. Enough of my brain's neuronal network for that person had survived to be activated by the cues on offer. But I could not recall his name. He had changed considerably from the 10-year-old boy I had known back in the old days. In addition, the network in my brain for my friend had atrophied somewhat. As a result of the brain's plasticity, as described in chapter 5, the neuronal networks that house our memories change over time and with use.

So, surely an image recognition algorithm with broadly the same architecture as the brain's cue-addressable memory ought to remember objects just as reliably as the brain. But it does not. Remember from chapter 5 the researchers who managed to trick such algorithms into mistaking a picture of a turtle for a rifle by tweaking some pixels. Whilst human brains can make mistakes too, they are more robust. Not all of the relevant (human) neuronal network's inputs are needed in order for the network to be activated. So, a few distorted or out of place pixels will make no difference. It also appears that human brains possess what is in effect a common-sense feedback mechanism. This relies on knowledge, including context, stored in the brain. If the brain's memory recall is about to misclassify an object, its common-sense mechanism notes the discrepancy between the potential misclassification and its stored knowledge. It can then feed its correction back into the recognition and recall process.[24]

How brains reason

Let us return to the brain's dual feature capability of knowledge plus processing. The two, knowledge and processing, go hand in hand. We know what Halle looks like; that is the knowledge. Recognising Halle in a crowd is the processing of that knowledge. Both features are located in the Halle neuronal network. In fact, the knowledge of Halle is one and the same as the processing of Halle by the Halle neuronal network.

This type of thinking is an example of what is called "System 1"[25] thinking or Type 1 processing. It belongs to a theory of how the brain thinks or reasons called dual process theory. Dual process theory was popularised by Daniel Kahneman in his widely-sold book, *Thinking, Fast and Slow*. System 1 thinking is, as you might imagine, the fast thinking of the title of the book. It "operates automatically and quickly, with little or no effort and no sense of voluntary control".[25]

System 1 thinking accounts for the great majority of our thinking. It can handle recognition of objects and faces. It can handle simple mental arithmetic and simple conversation. Most of our daily routines such as brushing our teeth or driving to the local shop are products of System 1 thinking. But precisely because it operates automatically, it can easily be led astray. The phenomenon of priming illustrates the point.[26] Kahneman gives the example of someone having recently heard the word "eat" being predisposed to complete *SO_P* as *SOUP* rather than *SOAP*. Conversely, having recently heard "wash", the predisposition goes the other way, completing the word as *SOAP* rather than *SOUP*. In each case, our brain has been primed or framed by hearing a prior word that automatically biases it in one direction or another. Accuracy and truth are not the be-all and end-all in System 1 thinking. But we need System 1 thinking, even if it is imperfect every so often, because it is so fast and therefore efficient.

The other prominent feature of System 1 thinking is that it is unconscious. Only once the thinking has been done do we become aware of it. For example, you become aware of the fact that you have recognised Halle's face in the crowd. The recognition suddenly appears in your mind.

But what if you cannot quite recognise her? This looks like another 'it's on the tip of my tongue' moment: you might have seen that face before but you are not sure. You need to think actively about it and then a minute later the answer appears. What you have just done is shift into *System 2* thinking, which is *effortful*. You have needed to tell yourself to concentrate. You might ask yourself questions such as where you might have previously encountered this person. It feels as if you might be doing the thinking yourself rather than your unconscious brain doing it for you.

System 2 thinking probably evolved well after System 1 thinking.[27] Most animals are perfectly capable of surviving in their traditional environments purely on the strength of System 1 thinking. It is also worth adding that, having used System 2 thinking to learn something, a skill for example, that skill can then be transferred to System 1. The skill in question, how to drive a car, for example, becomes automatic.

System 2 thinking appears to involve the new and different feature of symbol manipulation. It is as if the brain forms symbols or mental representations of the objects or parts of objects that it perceives. The brain can then pay attention to them and identify how they might be related to, or combined with, other symbols or mental representations that it already has stored away. These relationships are logged by the brain as one-on-one relationships but they can then be extrapolated to other situations. You might have seen Halle's face before but you are not sure. That smile stands out. Hmmm, where have I seen that before? Was she in a film? Yes, that was it; she was in that Bond film.[28] Her name is Halle Berry. That sort of thinking process is not statistical. It is step-by-step, or serial.

We will come back to the distinction between the probabilistic or statistical nature of System 1 and the symbolic nature of System 2 shortly. It is a distinction of fundamental importance to the creation of artificial general intelligence. It leads to the second of the Two Hurdles.

Kahneman attributes the terms System 1 and System 2 to two other psychologists, Keith Stanovich and Richard West,[29] although they called them Type 1 and Type 2 processing. Type 2 processing serves to suppress[30] Type 1 processing. If it did not, then we would generally slide back into Type 1 processing because it involves much less effort. Indeed, if we are feeling tired or weak-willed, that is exactly what happens anyway. We talk about 'lazy thinking'. The long war to prove the link between smoking and lung cancer is a classic example of lazy Type 1 thinking eventually succumbing to the cold deduction of Type 2 symbolic reasoning.[31] Daft though it might seem in hindsight, many attributed the increase in lung cancer in the mid-twentieth century to other reasons such as the growth in air pollution. This Type 1 consumer reasoning was comprehensively framed by the Type 2 advertising industry at the time. The advertising agencies successfully portrayed smoking as cool. People, smokers in particular, were perfectly happy to blame lung cancer on air pollution. It was only as a result of the development of a causal underpinning to the inferential reasoning that the link between smoking and lung cancer was conclusively proven.

Kahneman and Stanovich spend little time on the role of consciousness in their accounts of cognition, although Stanovich notes that Type 2 processing "is conscious rather than unconscious thought".[32] He does not develop the point about consciousness and what it might add. I believe that the exact role of consciousness is of fundamental importance in understanding how humans think and reason. It is the first of the Two Hurdles and is at the heart of developing the sentient robot. So, we need to work out exactly what role consciousness plays in thinking and reasoning.

Let us start by asking whether the idea of conscious thought makes any sense in practice. Ray Jackendoff, a linguist and a professor of philosophy, argues that all thought takes place in the unconscious mind. He writes that "I am inclined to believe that thought per se is *never* conscious".[33] So, when we think, our brain performs the thinking process. We then become conscious of our thoughts as they are translated into everyday language and appear to us consciously. Think about it, literally. We hear our own thinking as words and these words, of their own volition, appear in our heads in an unending stream.

Expressing the thinking process in this way implies that there is a language of thought that differs from everyday language. In other words, the brain performs thought in a medium that is compatible with the way that neurons work, not the way we speak to each other. As an analogy, one might think of the language of computation employed by a computer to perform numerical calculations. It does this in binary form, which lends itself to the on-off, or binary, nature of electric current. But it interacts with you, the user, in everyday numbers. Please hold on to this idea; we will return to it in the very next section.

To recapitulate, on Jackendoff's approach, the thinking that is present in both Type 1 processing and Type 2 processing is not conscious. We know that Type 2 processing takes effort and that, therefore, it is easy to slide back into Type 1 processing. That is suboptimal because there are many situations when Type 2 processing leads to the best answer. It looks as if we need a mechanism that prioritises Type 2 processing, which would otherwise not be activated very often due to the sheer effort involved. We encountered a similar problem in the previous chapter. I pointed out that the social emotions were, like Type 2 processing, valuable developments in the brain but prone to being overshadowed by the primary emotions.

In my view, the answer to the problem in both cases is consciousness. Consciousness does not perform the underlying thinking itself. Rather, I will argue that it plays the key role, albeit indirectly, in selecting Type 2 processing over Type 1. Type 2 processing is not always appropriate to the situation. We are all familiar with the expression, 'do not overthink things'. But, in the main, Type 2 processing is helpful. It needs to be supported against its default alternative, Type 1 processing. We shall consider this idea further in the next chapter and in much greater detail in part 3.

The language of thought

In the previous section, we started to look at the *hows* of thinking, ultimately landing on Type 1 and Type 2 processing. But we need to drill down a bit further. Do our *hows* of thinking survive collision with the knotty world of deduction, induction and abduction (hypothesis), as described in chapter 1, even if only at a conceptual level? This takes us back to the language of thought mentioned in the previous section. What is the language, or code, that Type 1 and Type 2 processing use?

Even though thought is unconscious, it seems intuitive to imagine that the language of thought in a human will be in English or Swahili, for example, depending upon the person's nationality. After all, we seem to experience thought or thinking in everyday words. But appearances can be deceptive. The thought process grinds away. Only then, as described, does the process become translated by the brain into the everyday words that we hear in our heads or indeed that come out of our mouths. The language of thought is not English or Swahili or any other everyday language.

To illustrate the point, let us begin to consider the language of thought as it pertains to a non-language-speaking animal such as a raven. In 2017, in a fascinating set of experiments, researchers showed that ravens can plan for both tool-use and bartering.[34] The results of the experiment were striking. The ravens clearly demonstrated the ability to use an unfamiliar tool in a novel task in order to secure a reward. Moreover, they could save the tool for a later time, contingent upon earning a later reward. They also learned to accept a token to exchange for the tool at the appropriate time. Perhaps most impressively, they could decide to wait for the later reward even when an alternative, immediate but smaller reward was available to them as they were making their decision. It is hard to see how one would not deem these mental abilities as thinking, even if they are non-verbal.

The ravens' success in the above experiment supports the idea that the brain possesses its own internal language of thought. After all, it is not as if ravens would employ English or Swahili or Hindi as their language of thought. Moreover, raven communications are highly unlikely to possess calls, or croaks, for things like tools, rewards and points in time. So, they cannot be thinking in 'raven' either.

Many experts in the field believe that everyday words are too vague to perform the logical operations that we need in order to generate coherent speech.[35] In other words, French and Mandarin, for example, are underpinned by an internal language of the brain. In order to think, the everyday words of our speech with others or of our own private dialogue need to be fed into a more robust format for (neuronal) processing. This internal format or language of the brain is sometimes called "mentalese".[36] Mentalese is not the same as everyday language. Instead, you can think of mentalese as the neuronal code that the brain uses to process thought, hence the description of it as the language of thought.

The alternative is to imagine that the brain uses everyday language to think rather than a special language of thought such as mentalese. On this analysis, the neurons in a German person's brain would form and directly manipulate German words in order to perform thinking. This idea is called linguistic determinism. It has now largely fallen out of favour, although it is widely accepted that each individual's everyday language has an influence on how they think.

Let us revert to the challenge of explaining how capabilities such as deduction, induction and abduction are realised in the brain. How exactly does the brain create a neuronal symbol for the letter G, for example? Just as importantly, how does the brain manipulate one set of symbols making up a set, such as books by John Grisham, so as to enable you to order them alphabetically in your bookcase after another set, such as books by Lee Child? Unfortunately, we do not yet know the answer. We know how a computer does it but we still do not know in detail[37] "how even the most basic computational operations might be carried out in neural 'wetware'".

Reading across from the workings of a computer to the workings of the brain is part of what is more generally known as the computational theory of mind (CTM). CTM was first developed by Hilary Putnam, a philosopher, in 1961 and subsequently elaborated by his student, Jerry Fodor. CTM neatly explains, solely at the conceptual level for now, how we humans can think logically and perform deduction. So, it seems to take care of aspects of Type 2 processing, albeit that it does not explain how effortful Type 2 processing is able to take precedence over easier Type 1 processing.

CTM represents a symbolic approach to the language of thought. It is particularly well suited to inferences made in the "realm of certainty"[38] such as the syllogism: all men are mortal; Socrates is a man; therefore, Socrates is mortal. The important aspect of this syllogism are the relationships between the words rather than the words themselves. In philosophical terms, the emphasis is on the syntax, or structure, of the syllogism rather than on the semantics, or meanings of the words, in it. We could instead say: all men are from Mars; Socrates is a man; therefore, Socrates is from Mars.

Symbolic approaches, however, "would be challenged to capture all the gradations of reasoning people find so intuitive and valuable in an uncertain world".[38] Most of the time, we are just not sure enough of the facts to reason with such logical certainty. This was the problem that gave rise to the catchphrase 'does not compute', used by the robot in the US TV show, *Lost in Space*.[39] To overcome that challenge, let us meet a theory competing with CTM called the "probabilistic language of thought" ('PLoT').[38]

PLoT, despite the name, does not throw the symbolic baby out with the bathwater. Rather, it seeks to draw on both the symbolic approach of CTM and also a statistical or probabilistic approach to inference. The latter approach deals more effectively with many of the uncertainties of the real world. Recognition of patterns, images and objects falls more naturally to a statistical inferential mechanism. Let us take Bertrand Russell's turkey,[40] for example. Each morning, he wakes and eagerly anticipates his breakfast. He is fed at 9.00 a.m. every day. He has successfully induced the conclusion that the very fact of his waking up causes the appearance of breakfast. Indeed, whenever we (automatically and without effort) induce a conclusion on the basis of our observations, we are doing so on the basis of probabilities or statistics. This is Type 1 processing at work.

When a human being perceives a steady correlation, lazy Type 1 processing may lead to the conclusion that there is a causal relationship between the two events in question. In fact, the causal relationship may or may not hold true. Type 2 processing, if activated, helps to challenge Type 1 processing. In this case, a turkey with a humanlike brain would probably ask himself why the great philosopher was feeding him such a good breakfast each day. This exceptionally clever turkey with the highly unusual brain would conclude that the correlation was misleading. His mere awakening does not cause his breakfast to arrive. In fact, there is a different causal relationship altogether. Russell is feeding him in order to fatten him up for eating. Breakfast just happens to be a good time to feed farm animals.

Indeed, the day before Christmas, the turkey (the normal one, without a humanlike brain) awakes and his throat is unceremoniously cut. If only the turkey had been able to ask himself the question, why are they putting in all this effort to feed me every day? His problem was that he did not possess Type 2 processing abilities. He could not take a symbolic approach to thinking.

The statistical approach does not allow us to combine what we have learned into new thoughts or ideas. So, whilst I can use Type 1 processing to recognise Halle, it is not going to help me work out how to persuade her to go on a date with me. That is a new idea altogether, and probably to the reader a rather surprising one. This capability for manipulating different symbols or mental representations of objects in the world, in this case, Halle, me and dinner, is called "compositionality" in the jargon.[41] The symbolic approach is needed to enable concepts to "be flexibly combined with other concepts to form an infinite array of thoughts in order to reason productively about an infinity of situations".[38] So, PLoT incorporates a symbolic capability as well. It gives its possessor the ability to perform both Type 1 and Type 2 processing.

Both CTM and PLoT have drawbacks. We do not understand how the brain performs cognitive computations at the neuronal level. We are still very much at the conceptual stage. Secondly, the two theories do not explain how the human brain performs abduction. This is the ability to hypothesise in the absence of many of the steps in the thinking process and with only a few data points to hand. Albert Einstein's theories of relativity were hypotheses. Hypothesis, or abduction, is responsible for the great majority of mankind's advances over the millennia. As Fodor noted, "Abduction really is a terrible problem for cognitive science, one that is unlikely to be solved by any kind of theory we have heard of so far."[42]

Herbert Roitblat describes abductive challenges as "insight problems".[43] He notes that they "are often associated with a subjective feeling of 'Aha', as the solution is discovered". Solving such problems relies on imagination and creativity, which we consider further below. Put another way, solving such problems relies on being able to look at, or represent, them through another lens. You can spend hours wrestling with a problem and getting nowhere. You might then come back to it and the problem literally re-presents itself to you. You see the problem in a different way.

This ability relies in turn on having a model of the world in your brain. The trouble is that we do not know, at beyond a conceptual level, what a model of the world looks like inside the brain. It will be instantiated in the form of countless millions of neuronal networks with an extraordinarily high level of interconnectivity. In that way, one idea or perception or connection can spark another idea seemingly out of the blue. At this stage, though, that is about as far as we can go. Indeed, it is not even clear that abduction falls into either Type 1 or Type 2 processing. Perhaps it comes as blend of the two.

Our lack of understanding does not just extend to the big insight problems. It extends too to the little insight problems, otherwise known as common sense. Common sense also relies on having a so-called model of the world built into your brain. Somehow or other, human brains know what looks right in the environment around them. We talk about people having presentiments about a disaster that happens shortly thereafter. When asked why, they might respond, 'something didn't feel right.' In the technical jargon, that valuable ability is attributed to possessing a model of the world in your head.

This is not just a problem for cognitive neuroscience.[44] AI has the same problem. Much useful daily inference by humans, as well as the big theoretical breakthroughs, relies on having an enormous background store of richly categorised knowledge, a frame, so to speak. This is the model of the world in your brain. We do not yet fully understand how to load AI with a large body of knowledge such that AI can use it to perform abduction and common sense.[45]

We shall come back to the role of consciousness in cognition in chapter 10.

AI's take on cognition

AI's challenge then is to create a computational paradigm that can perform deduction, induction and abduction, in other words, a master algorithm. Only then can AI step up to becoming AGI. The leading lights of the deep learning world remain convinced that deep learning contains the answer.[46] They believe that in the fullness of time deep learning will be able to perform Type 2 processing. In addition, it will be able to learn an appropriately rich model of the world.

These beliefs derive from the fact that brains are currently the only systems in the world that exhibit general intelligence. So, there is an argument that AI ought to possess the same basic architecture as a (human) brain in order to achieve the same capability. Faith that deep learning's architecture will ultimately lead to emulating the human brain stems, in part, from the observation that the brain's cortex is "very, very uniform all over".[47] This might suggest that its artificial equivalent should be the same. It should be noted, though, that there are many different types of neuron and the depth of each of the layers varies from region to region. So, the brain is not as uniform as it looks.

Deep learning has enjoyed phenomenal success over the last few years. From image recognition to natural language processing to games-playing, deep learning has given us astonishing progress in a short period of time. But, critically, it remains largely stuck at the Type 1 processing level.[48] The big question for AI, which is arousing considerable sound and fury, is about how to combine Type 1 and Type 2 processing. There is a further twist to the tale. In order to make full use of the advantages of Type 2 processing, the requisite master algorithm also needs to contain an extensive body of knowledge about, or model of, the world.

Yoshua Bengio believes that the answer to Type 2 processing may lie in the development of machine consciousness.[48] Incidentally, this form of machine consciousness would not duplicate the subjective experience inherent in human consciousness. Bengio proposes techniques to lift generalisations out of the underlying sea of data. This would be done by focussing attention (hence the allusion to consciousness) on the relevant generalisations. These generalisations would then be transferred into, and manipulated within, other domains of thought, thereby mimicking the symbolic features of Type 2 processing. This digital *legerdemain* would all be performed within the ANN paradigm, specifically in deep learning. Put another way, the parallel nature of ANNs would become translated into the serial nature of logical thinking and reasoning.

Maybe the ability to perform abduction would follow. To repeat the point, performing abduction often involves seeing a problem in a new way or in a new representation. Currently, algorithms do not choose the way that they look at problems. The choice of representation used by an algorithm, i.e., the way that an algorithm looks at a problem, has been preselected by the algorithm's designer. It is part of the algorithm's design. As Roitblat observes,[49] "a truly general computational intelligence will require the ability to create its own representations." This means that AGI must be able to step back and then approach difficult problems in a wholly new way in order to crack them.

A competing school of thought in AI asserts that deep learning will not be able to take the strain of Type 2 processing and building a rich model of the world on its own. It proposes that methods must be found to weave the statistical approach together with the symbolic approach in a hybrid master algorithm. It even has a name, neuro-symbolic AI.[50] Quite apart from the challenges of weaving the two approaches together, there is the challenge of developing a rich model of the world in the first place.

There are just a few initiatives tackling this challenge. I mentioned Cyc in chapter 1. Diffbot is another such initiative.[51] Diffbot effectively reads the whole of the worldwide web every few days so that its knowledge base is always as up to date as possible. Critically, it then creates relationships between all those individual facts that it has hoovered up. So, similar to the way our brains have created an enduring relationship between Halle Berry and that Bond film, so can Diffbot do likewise. It is able to create an extraordinarily voluminous web of relationships between all the things in the world that have been represented on the worldwide web, in other words a rich model of the world. Such a model can be used to underpin symbolic reasoning and abduction.

There is a further dimension to the challenge of building a world model, as mentioned in chapter 5. This is the notion of embodied cognition. A substantial amount of what we learn about the world as humans is achieved by virtue of our bodies and, in particular, our five main senses. At its simplest level, upon encountering an object the brain is able to use the feedback from each of the five senses to learn about, and record, this object in multiple ways. In future encounters, the brain is then able to cross-check what it perceives in some or all of these different ways. It is becoming clearer that, in order to build a world model, AI might need to employ some or all of the same sensory capabilities, albeit in digital form. For example, researchers have discovered that a robot's object recognition is improved by also being able to listen to the sounds made by the object in question.[52] So, the robot does not need to rely on sight alone. We will revert to this topic in part 5, 'Building the Sentient Robot'.

In order to emulate Type 2 processing, we might also need to devolve some of the work to AI. That means developing AI that can itself write code. This is a different task to the self-coding that machine learning algorithms effectively perform during training. The latter takes place within a pre-established architecture. The former is about developing AI that writes its own architecture. DeepMind has recently announced the creation of AlphaCode to do exactly this type of job.[53] As the AlphaCode team observes: "For artificial intelligence to help humanity, our systems need to be able to develop problem-solving capabilities." AlphaCode still has a long way to go but, in coding competitions hosted by Codeforces,[54] AlphaCode ranked roughly in the middle of the (human) entrants.

Whichever of the two schools of thought is correct, deep learning on its own or a hybrid approach, solving the problem is of fundamental importance to the building of artificial general intelligence. This is the second of the Two Hurdles.

Chapter 8

Decision-Making

Types of decision-making

When you consider how and when you make decisions, you probably think of weighing up the options, giving the process due time and consideration. Decision is a word that suggests some level of careful thought. We talk in unflattering terms of someone making hasty decisions. Actually, most decision-making is not considered or measured at all. The basic level of decision-making is about making sense of what we perceive. This is termed perceptual decision-making.

In the previous chapter, I described how the brain was able to recognise Halle in a crowd as a result of the activation of the relevant neuronal network. The network in question was effectively purpose-built for Halle recognition. The result of learning how to recognise Halle for the first time was the creation of a Halle neuronal network. So, subsequently, when Halle was spotted, the sight of her gave rise to certain patterns of neuronal activity. These in turn triggered the beginnings of the Halle network. A domino effect rippled through the network and the result, recognising Halle, happened as if automatically. Even though the process is rapid and automatic, this process is called perceptual decision-making.

The more advanced form of decision-making is termed value-based decision-making (VBDM).[1] This is the type of decision-making that entails a comparison of competing options based upon a mental evaluation of each of them. VBDM can be broken down into two further buckets: stimulus-response and goal-directed[2] (see figure 8.1). Stimulus-response decisions tend to be simpler and habitual and the decision-making process automatic, even if potentially sophisticated. Goal-directed decisions tend to apply in novel situations or involve autonomously set goals.

Figure 8.1
A taxonomy of decision-making

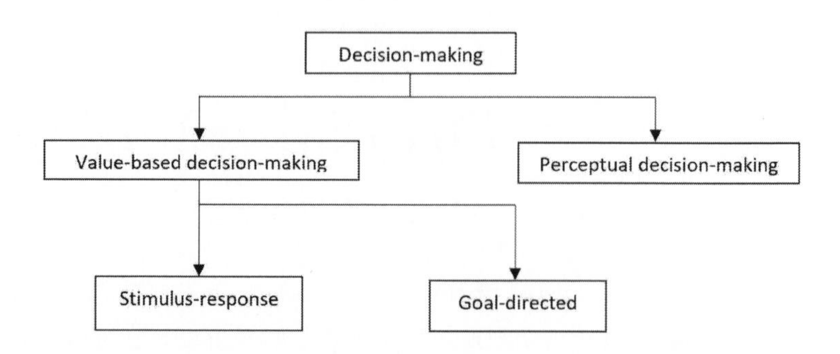

Nathaniel Daw, a professor of computational and theoretical neuroscience, and John O'Doherty, a professor of psychology, describe research work on decision-making involving two groups of rats. Each group of rats was trained to press a lever to secure food.[3,4,5,6,7] One group was moderately trained whilst the other was intensively trained. Both groups of rats were then free to press their levers and eat until fully sated. At that point, the levers were taken away and the rats sat there to begin digestion. The levers were then presented to the rats again while they were still full.

The intensively trained rats reverted to pressing the lever and eating despite already being full. This was stimulus-response VBDM at work. As we shall see, the underlying neuroscience is quite sophisticated but the outcome is automatic. In contrast, the moderately trained rats desisted from the temptation to over-eat. They retained their own autonomy, their own goal-directedness. Their goal was to eat well when the opportunity, i.e., the lever, presented itself, but no more than that. This was goal-directed VBDM.

Further research suggests that humans are susceptible to the same sorts of influences on their behaviour as the rats.[8] This yields the same distinction between stimulus-response and goal-directed behaviours. They work in different ways but underpinning both is an elegant blend of emotion and cognition. In simple terms, VBDM takes place over three phases in the brain. These phases roughly equate to the sequence: evaluate, decide, learn.

Value-based decision-making; phase 1

In phase 1, evaluate, alternative courses of action are compiled and each assigned a value. The value is expressed in the form of neurons firing. This is a complex process, mediated by different neuronal pathways. These come together incrementally. Inputs to this evaluation include cognitive activity, remembered utility (the anticipated rewards from each possible action, as informed by memory, i.e., past experience), emotional inclination, emotional context, assessment of riskiness, assessment of time-to-reward and costs of action. Every single one of these inputs is worked out by the unconscious brain, or mind. Importantly, the brain is also able to place a value on an outcome that is deferred. It does this by discounting the value of the deferred reward to compensate for the fact that it is not immediate.[9] Remember the ravens. They were able to decide to wait for a later reward even when an alternative, immediate but smaller reward was available to them.

As a result of all this neuronal computation, each possible action winds up with a net action value in the brain. This net action value is essentially the predicted reward from the action, discounted for how far in the future it is, minus the predicted costs to achieve it.

The parts of the brain that contribute to decision-making are numerous. This is not surprising given the number and type of inputs to this first evaluation phase, as described above. In general terms, it appears that the ventromedial prefrontal cortex, the orbitofrontal cortex and the ventral striatum compute the subjective valuations of choices (see figure 8.2).[10] In contrast,[11,12] the dorsolateral prefrontal cortex appears to moderate the conclusions from certain of these computations of value. Indeed, the dorsolateral prefrontal cortex is thought to play a role in self-control. The anterior cingulate cortex also plays a part. This is especially with respect to evaluating or costing the effort involved in a possible particular course of action.[13] Recent research has also implicated the cerebellum, the cauliflower-shaped part of the brain sitting low down at the back (see figure 6.1).[14,15,16]

Figure 8.2
Evaluating regions
of the brain

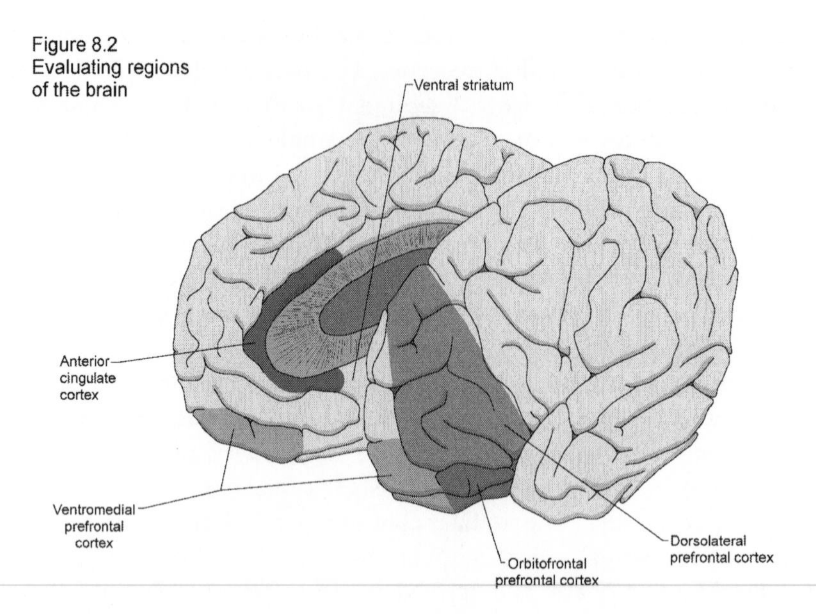

It is especially important to remember the emotional input to this mechanism. This is not just an exercise in cognition; rather, it is an exercise in combining cognition and emotion in order to weigh up behavioural alternatives. This important point was captured by Antonio Damasio in what he called the "somatic-marker hypothesis", as cited on p. 32 in chapter 2. Indeed, decision-making reaches all the way down to parts of the brain such as the amygdala,[17] deep in the limbic system.

Value-based decision-making; phase 2

In phase 2, decide, the alternative courses of action, having been evaluated, are compared. One course of action then comes to dominate. This all takes place at the neuronal level. This is not a case of static ordering or ranking of the alternatives. Rather, the process at work is termed stochastic. The *Oxford English Dictionary* defines stochastic as "having a random probability distribution or pattern that can be analysed statistically but not predicted precisely".[18] In simple terms, one cannot exactly predict which of the alternative courses of action might emerge at any given decision point—the decision is random. But, given, say, 100 repetitions of a particular decision, one can estimate the probability in each case of the brain going one way or the other.

The process by which this happens is once again rooted in neuronal networks. Paul Glimcher, a professor of neuroeconomics, neuroscience and psychology, terms the process, "winner-take-all computation"[19] (see figure 8.3). (Neuroeconomics is the study of decision-making in the brain.)

The alternative net action values are expressed in persistently changing terms. This is both because the neuronal inputs themselves bounce around and also because each alternative excites or inhibits its competitors. Nonetheless, since it is up to the brain to generate behaviour, which means a single behaviour, a participating or competing alternative finally wins the contest; hence the term, winner-takes-all.

Figure 8.3
Winner-takes-all computation

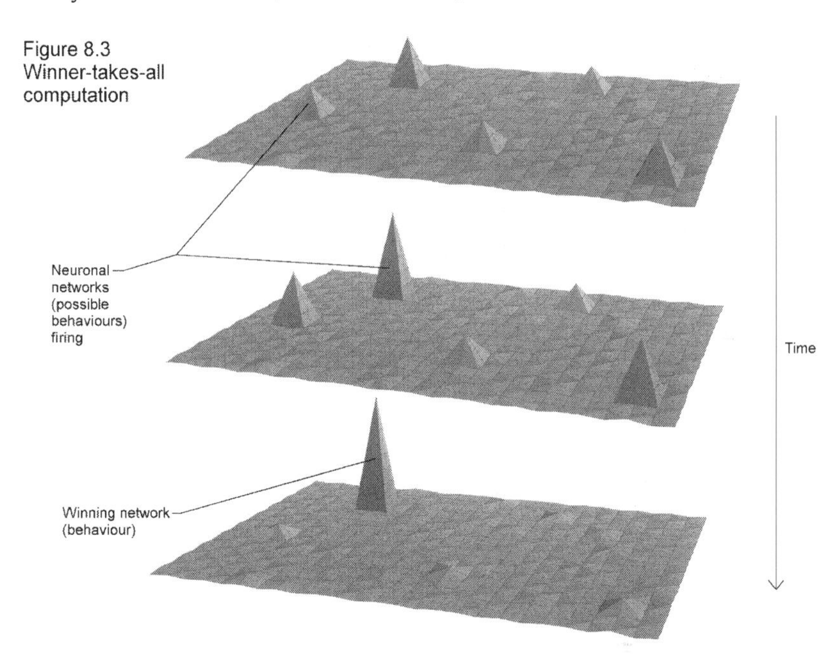

The way that the brain performs such computations is still unclear, although computational neuroscientists have a number of possible alternative models. Once the winning behaviour has been selected, the choice is projected to the motor cortex, thence to the spinal cord and finally to the relevant muscles. Behaviour has been generated.

It may be thought strange that phase 2 of decision-making is probabilistic. This way of making decisions is an outcome of the way in which autoassociation networks operate, reliant as they are on their backprojections. Rolls describes[20] this probabilistic feature as potentially evolutionarily advantageous since it:

> ...may contribute to learning. Consider for example a probabilistic decision task in which choice 1 provides rewards on 80% of the occasions, and choice 2 on 20% of the occasions. A deterministic system with knowledge of the previous reinforcement history [experience] would always make choice 1. But this is not how animals including humans behave. ... By making the less favoured choice sometimes, the organism can keep obtaining evidence on whether the environment is changing...

In other words, the human decision-making system is random but not chaotic. There is a method to the madness. Choice 2 may turn out to produce a better outcome than choice 1 in certain circumstances. The organism thereby learns to promote choice 2 from time to time.

Interestingly, it does not appear at first sight that our brain's ability to come up with different behavioural options, evaluate them and then generate a winning behaviour needs consciousness in order to work. We seem to have an explanation for all the moving parts in decision-making without having to turn to consciousness. That is of course not to say that we are not conscious of our own decision-making, or at least of the final decision. Our explanation makes the processing in question sound like the workings of a computer. It is certainly computational, albeit not in the same way as a standard computer operates. One should instead picture different, competing neuronal networks. They each fire at their own rate and strength, calibrated by relevant inputs. As we shall see later, I argue that consciousness does in fact play a part but in a rather unexpected way.

Value-based decision-making; phase 3

In phase 3, learn, the outcome of the action or behaviour is itself evaluated, learned and stored as remembered utility for use in phase 1 of another decision-making cycle in the future. Remembered utility consists of the anticipated rewards from each possible action, as informed by memory, i.e., past experience.

Phase 3 is more contentious in that there is more than one theory for its operation. A leading hypothesis explaining the learning mechanism is the dopaminergic reward prediction error ('DRPE') hypothesis. Let us take a novel and pleasant sensation. Given the brain's ability to learn quickly, a memory of the sensation itself and the action taken to achieve it will be stored. This means that, upon the perception of the same circumstances on a future occasion, the brain can predict that it will receive the same neurotransmitter rewards if it pursues the same action. Note that these rewards will not be expressed in dopamine, since dopamine is released in respect of wanting rather than liking, as described in chapter 6. The prediction will be one of the inputs in phase 1 of the decision-making process, as set out above. If the brain then goes on to decide to take the relevant action and if the rewards are greater than predicted, then the brain will have made an error in its original prediction. That error will be expressed in terms of dopamine. In other words, dopamine is concerned with the error, not the reward *per se*.[21] The identification of the error will cause the brain's memory of the sensation and the action taken to achieve it to be adjusted, in this case positively. In future, we will want the relevant stimulus even more than we did previously. This adjustment happens by way of changes in synaptic strengths. Thus, the brain is constantly updating its memories to reflect its latest experiences.

Due to its basis in a reward system, the DRPE mechanism bears an uncanny resemblance to reinforcement learning in the world of AI. I first mentioned reinforcement learning in chapter 1 as one of the means by which machine learning algorithms can be trained. In a machine, the dopaminergic and other neurotransmitter (such as opioid) pathways are replaced by digital reward functions. Indeed, this is an example of the two worlds of AI and neuroscience being very similar to each other.

Many choices, however, are not resolved with one behavioural act. During a rally in a tennis match, I am having to make decision after decision, one shot after another, deciding where to place the ball, how hard, with what spin, etc. Until the rally has ended, the real reward, winning the rally, is not available. No intermediate individual shot confers that reward. That reward only comes at the end of the rally, and only if I have won it. How does the brain optimise all those sequential decisions? The answer lies in a type of reinforcement learning called temporal difference learning. Temporal difference learning relies on prediction error, as just described.[22,23,24]

Let us say that a sequence of three shots on your part will win you the point. The third shot, the winning shot, should have the highest reward attached to it: it is the winning shot, after all. The second shot needs to put your opponent under the right amount of pressure such that his or her return enables you to hit the winning shot. So, that second shot needs to have quite a high reward attached to it, but definitely not as high as the real reward attached to the winning shot. The first shot puts the whole play into motion. Accordingly, the first shot has a reward attached to it, but not as high as the reward attached to the second shot. In effect, the anticipated reward at each decision point (i.e., each shot) is borrowed, or brought forward, from the reward built up cumulatively during the rally but only experienced at the end of the rally.

Models of the world

The reinforcement learning described so far, including temporal difference learning, is what is called model-free learning. Daw and O'Doherty equate model-free learning with stimulus-response learning (see figure 8.1).[25] The trouble with stimulus-response learning is that it "is incapable of evaluating novel actions (or re-evaluating previously experienced ones) based on any other information about the task, the world, or the animal's goals".[26] Stimulus-response learning only works when triggered by the same inputs that were triggered previously. It cannot take into account new information.

Computationally speaking, model-free learning is narrow in scope. It relies purely on possessing an updateable capacity. This capacity can inform the brain's decision-making mechanisms as to the reward for such and such an action. The brain does not need to possess a model of the surrounding environment, a model of the world, for such a capacity to function. Hence, it is model-free.

Machine learning today generally operates in a model-free way. Take the example of a so-called convolutional neural network (CNN) with a skill in mouse-recognition, for example. It does not need to have a model or a view of the world to recognise mice. All it does is recognise mice; it is model-free. Or take a self-driving car. It only needs to perceive objects that might immediately impact or enable its progress. It will log, in a simplistic way, the large office block on its left but it will not perceive the receptionist seated behind the desk in the ground-floor lobby.

As Daw and O'Doherty point out, such an algorithm cannot evaluate novel actions. It is also unable to take into account more complex goals that the animal might have. For that, we need to turn to model-based reinforcement learning.[25] Model-based learning relies on mental representations, or models, of the surrounding environment and possible outcomes or futures. The brain needs to have a mental picture of what things will be like if it decides to take one action; and it needs another mental picture of what things will be like if it decides to take some other action. It needs to be able to simulate these possible outcomes. Each of these model futures then needs to be evaluated, so that comparison is able to take place and a decision or choice made between them. Model-free learning does not make such choices on the back of such simulations or model futures. Rather, it just looks at what rewards were generated in the past from each possible action. Model-free learning is automatic by comparison with model-based learning.

As an aside, let us briefly note an interesting development in the thinking around AI. It is increasingly believed that model-based reasoning techniques will need to be at the core of a true thinking machine.[27] But researchers also note that "there is abundant evidence that model-free mechanisms play a pervasive role in human learning and decision-making (Kahneman, 2011)."[27,28] In short, humans display both model-free and model-based learning and reasoning. Unsurprisingly, it looks as though thinking machines will do the same.

We have come back to the same place as we reached in the previous chapter. There, I described Type 1 and Type 2 processing, concepts that arise from the world of psychology. The account of the different inputs to decision-making described immediately above, model-free and model-based learning, arise from the neuroscientific world. But the two accounts, from the worlds of cognitive psychology and neuroeconomics, respectively, come up against the same issue, the second of the Two Hurdles. We know what the brain does when it uses its rich model of the world to engage in Type 2 processing and take goal-directed decisions. But we do not yet know how it does it. The world of AI is in the same situation.[29] At least we know what we do not know. That is always an important step on the road to the answer.

Decision-making in practice (I)

Up until now, when we have considered cognition, we have looked mainly at how it works, or might work. This exercise led us to the distinction between Type 1 and Type 2 processing. Similarly, when we have considered the emotions, we have looked at how they work. That exercise led us to the distinction between the primary and the social emotions. We now need to consider *whether* they work rather than how they work. This means, specifically, whether the mechanisms described so far are conducive to the smooth and effective operation of Type 2 processing and the social emotions, respectively. For these latter two facilities are the latecomers to the party. Type 1 processing and the primary emotions have both been in existence for much longer. They require less energy in the sense of effortful brain power. Behaviours underpinned by them should slide easily through the decision-making process without giving behaviours underpinned by Type 2 processing and the social emotions a look-in. That would not make sense given the self-evident utility of both Type 2 processing and the social emotions.

Daw and O'Doherty ask this question in the sense of trying to identify the mechanism the brain might use to determine whether to select a goal-directed decision, underpinned by Type 2, model-based cognition, or a sensory-response decision. They observe that:[30]

> One possibility is that a separate arbitrator or controller sits on top of these systems and acts to gate their access to behaviour. Alternatively, the systems might somehow competitively interact without the need for a separate arbitrator, e.g., through some sort of mutual inhibition at the point of action selection. Empirical evidence in favor of either of these possibilities is currently lacking.

A "separate arbitrator or controller" sounds suspiciously like the dreaded *homunculus*. The homunculus is a philosophical *bête noire*. In an effort to explain the feeling that there is an 'I' sitting inside my head directing events, the homunculus theory suggests that a homunculus ("a very small human or humanoid creature")[31] is sitting inside my brain. The objection to the homunculus theory is that there would have to be another homunculus, a second one, sitting inside the head of the first homunculus.

Otherwise, how would the first homunculus work? But then there would have to be yet another homunculus, a third one, sitting inside the head of the second homunculus. Otherwise, how would the second homunculus work? This is *reductio ad absurdum*. In other words, the theory leads ineluctably to an absurd conclusion.

Anyway, we do not need a separate arbitrator or controller. Daw's and O'Doherty's competitive interaction approach is the stochastic winner-takes-all approach. What we need to strengthen Type 2 processing and the social emotions, respectively, is a helping hand during the winner-takes-all computation (see figure 8.3). Rolls proposes just such a helping hand in the form of a biasing mechanism acting on the relevant neuronal networks.[32]

When the goal-directed approach first evolved in the brain, it probably came into play infrequently. How would that have worked? The brain might have paired its newly acquired goal-directed approach with a mechanism that subtly casts doubt on a potential stimulus-response decision in certain circumstances. Rolls proposes such a mechanism.[33] He proposes that the brain evolved a method whereby it could rate the confidence with which it made a decision. If the level of confidence in that decision were not high enough, then the decision-making network could be impelled to start again. He proposes that the mechanism for this is simply "a separate confidence decision network", with reciprocal connections to the main decision-making network. He notes that a network of this nature is "computationally simple".[34]

So, in a certain situation, Type 1 processing might generate an answer to a problem. The separate confidence decision network might, however, give the answer the thumbs-down. In such a case, the decision-making network would run again. This would afford Type 2 processing another opportunity to advance its solution to the problem. Type 2 processing, Kahneman's "thinking slow", might therefore triumph, but only infrequently. For most animals lucky enough to possess a measure of Type 2 processing capability, that might be enough.

But the brains of some animals continued to evolve in power and complexity, as did the societies they inhabited. I suggest that millions of years ago a small number of animals, perhaps the forerunners of today's great apes, elephants, corvids and certain cetaceans, evolved a yet more effective biasing mechanism. This mechanism really drove goal-directed decisions forwards to the point at which they could support and sustain the complex societies that such animals were forming.

Decision-making in practice (II)

Type 2 processing and the social emotions both have real utility in our society. It would be implausible that evolution, having furnished us with them, did not enable us to make use of them more than just every so often.

Recall that the nature of decision-making is stochastic. I believe that a purely stochastic framework for decision-making only goes so far. Our social systems, the way we live with other people, are so complex and have to be so nuanced that purely stochastic decision-making on its own would be too uncertain to be stable and sustainable. It might have worked millions of years ago but human society has come on a long way since then.

Mankind has developed an extraordinary range of social emotions, as described in chapter 6. Such emotions need to be deployed with a fair degree of predictability. Only in that way might other people, such as family, friends and colleagues, rely on them and ultimately on each of us. Purely stochastic mechanisms for selecting as between a primary emotion behaviour and a social emotion behaviour would be insufficiently reliable.

It would be reasonable to assume, for example, that sympathy is a social emotion that people are expected to experience, and thence exhibit, in what society might deem to be appropriate circumstances. In many cases, the competing emotion, for example, might be a primary emotion such as anger or fear. Picture being asked by a lost-looking stranger in the street for directions. Most of us react sympathetically. What if anger and fear, which are primary emotions and are by definition stronger and more commanding than social emotions, probabilistically prevailed? Visiting foreign cities on holiday would become a nightmare.

I suggest that the stochastic framework requires a biasing input to achieve the reliability described above. We would also need such a biasing input to give Type 2 processing time to work. When confronted with a problem, the temptation is to go with the first answer that pops into your mind. That will be the Type 1 answer. Coming up with the Type 2 answer takes effort and, if the problem is tough enough, plenty of time. Picture yourself grappling with such a problem. You will experience moments when you are just about to give up. But your strength of will forces you to keep going, to keep thinking about the problem. That is the biasing input at work again, effectively putting the Type 1 answer back in its box in order to allow Type 2 processing to keep working at the problem.

I propose that consciousness is the biasing input in both cases. It evolved to enable its possessors to help address the greater complexity of the advanced social groups in which great apes and certain cetaceans came to live. It also evolved to sustain Type 2 processing to help tackle complex novel decisions. Rolls too concludes that such decisions require consciousness. This leads him to his "higher order syntactic theory (HOST) of consciousness",[34] which I describe in more detail in chapter 11. My own theory of consciousness springs from a similar analysis but then adopts a somewhat different path, as I set out in part 4.

In summary then, cognition, emotion, decision-making and the nature of the society that each animal inhabits are closely entwined (see figure 8.4).

Figure 8.4

Progressive complexity in decision-making reflected in types of animal and their societies

Cognition	Emotions	VBDM	Types of animal	Social interaction
Type 1	Primary	Stimulus-response	Most	Simple or regimented
Type 1; simple Type 2	Primary; simple social	Stimulus-response; simple goal-directed	Primates; higher mammals; some birds	Advanced
Type 1; complex Type 2	Primary; complex social	Stimulus-response; complex goal-directed	Great apes*; elephants; corvids and certain cetaceans	Sophisticated

* includes *Homo sapiens*

I should note, for completeness, that it is often assumed that, since evolution definitionally leads to fitter outcomes, it also leads to better outcomes. Not all social emotions are good, especially if taken to extremes. Overweening pride and a thirst for dominance can give rise to dictatorships. On this analysis, if consciousness enabled such emotions to outcompete the primary emotions, then the resultant social system might be stable and sustainable but it would also be unpleasant. In his book *The Better Angels of Our Nature*,[35] Steven Pinker, a professor of psychology and a linguist, sets out a rosy view of human nature. Over time, violence in human society has fallen and cooperation has grown. But that does not necessarily lead to perfect outcomes—politics still throws up the occasional Stalin or Mao, for example—just more stable ones.

Chapter 9

Summary of Key Points from Part 2

Part 2 has explored the workings of the brain with particular attention to the emotions, cognition and decision-making. This will serve as a basis from which to explore the topic of consciousness in detail.

1. Living organisms have evolved with increasingly complex emotional and cognitive neuronal circuitry. Different and competing behavioural drives emerge in such organisms. The brain has evolved a means of prioritising amongst competing potential behaviours, a decision-making system.
2. The brain is composed of different types of cells. The most important are neurons. Each neuron is able to 'fire' as a result of sufficient stimulation from neurons connected to it. The neurons at the beginning of each chain, so to speak, are stimulated by internal or external sensory inputs. All neurons are connected to thousands of other neurons. Such connectivity enables neurons to organise themselves into neuronal networks. The ability of neurons to 'fire' combined with their network connectivity turns the brain into an information processing platform.
3. The information processing that takes place in between the brain's sensory and motor functions is what is of interest to us in this book. In the simplest organisms, such as a bacterium, the order of events is effectively sense-move. In *Homo sapiens*, the order is sense-process-move, with the emphasis on 'process'.
4. Some of the most important neuronal networks with respect to process are the decision-making networks. These networks receive inputs both from the brain's neuronal pathways used in perceiving and registering emotions and also from its cognitive capacities.

5. Emotions are instantiated in the brain in different systems or constellations of neuronal networks. We understand the brain's primary emotions better than we understand the brain's social emotions. The systems for the primary emotions originate in the ancient mid-brain and project all the way to the much more recent prefrontal cortex. In this latter location, the emotions feed directly into the decision-making networks.
6. The social emotions feed into the decision-making networks too. The winner-takes-all enabling mechanism that allows the brain to prioritise the relevant emotion in any given situation will, all other things being equal, tend to select the primary emotions over the social emotions. The primary emotions are more ancient with more extensive circuitry and are more powerful. But there is a drawback. Society could not function if primary emotions were generally prioritised over the social emotions. Humans, and perhaps certain other animals such as elephants, must have evolved a further mechanism to reinforce the social emotions.
7. If emotion tells you where to go, cognition tells you how to get there. In the world of neuroscience, cognition is the process by which one comes to know something as well as the process by which one comes to arrive at a conclusion given what you know. In thinking about how cognition operates, we need an account that addresses both how the brain acquires and uses knowledge in habitual circumstances and also how the brain reasons in new situations. The former relies heavily on learning and memory. The latter relies on deduction, induction and abduction (hypothesis).
8. The brain appears to possess two ways of performing cognition, of processing information. These are known colloquially as thinking fast and thinking slow. Thinking fast engages Type 1 processing. This is quick, automatic and probabilistic. Thinking slow engages Type 2 processing. This is slower, effortful, more energy-consuming and symbolic in nature. The brain engages in Type 1 processing to recognise a face, for example. It engages in Type 2 processing when it uses common sense to solve a problem.
9. The world of AI has mastered probabilistic Type 1 processing. It has not managed to emulate Type 2 processing other than in highly circumscribed formats. It still lacks common sense, for

example. This is the second of the Two Hurdles on the road to producing AGI.

10. The decision-making network is primarily formed of smaller networks located, variously, in the ventromedial prefrontal cortex, the orbitofrontal cortex, the ventral striatum, the dorsolateral prefrontal cortex and the anterior cingulate cortex.
11. The decision-making network operates in three phases. In phase 1, the network evaluates all the inputs to a decision, giving each of the possible competing behaviours a net action value. The net action value is essentially the predicted reward from the action, discounted for how far in the future it is, minus the predicted costs to achieve it. In phase 2, the alternative courses of action, having been evaluated, are compared using a winner-takes-all computation. The winning behaviour is then projected to the motor cortices of the brain for execution. The decision-making process is stochastic.
12. The decision-making network also provides for learning. This is phase 3. A leading theory says that the network predicts the net action value prior to action. The actual net action value, i.e., as a result of initiating the selected behaviour, is then compared to the predicted net action value by way of the so-called dopaminergic reward prediction error system. Any difference between the two (the predicted and the actual) has the effect of updating the brain for what is now the latest and best predicted net action value. This updated net action value will then be used in the next decision-making process involving the relevant behaviour.
13. Most Type 1 processing is model-free. By this, it is meant that the network does not need to invoke a model of the environment in which the decision is being made. In other words, it needs no context for making the decision. Type 2 processing is model-based. It relies on the brain possessing a model of the world around it.
14. Our social systems, the way we live with other people, are so complex and have to be so nuanced that purely stochastic decision-making on its own would be too risky and unpredictable to be sustainable. People's social emotions would translate into actual behavioural outcomes too infrequently. Moreover, Type 2

processing needs to be deployed often enough to derive its full benefit. Type 1 processing has its limits.

15. In order to address these problems, I propose that a mechanism has evolved in the brain to promote, or lend extra weight to, both the social emotions and to Type 2 processing so that they are deployed more often and more predictably. I argue that this mechanism is consciousness. It biases the stochastic model in the brain to minimise what would otherwise appear to be impulsive or random decision-making. Consciousness exists to shade the odds in favour of more predictable, more reliable behaviour.

Part 3.

Reflections on Consciousness

Chapter 10

Language, Thought and Consciousness

The holy trinity

To borrow a religious term, there is a holy trinity of capabilities that is fundamental to the generation of complex behaviour. The trinity comprises language, thought and consciousness. Whilst these three have distinct roles, they are so entangled with each other that they are often seen as one. Consequently, their individual roles become lost. We must first disentangle them. Only then can we address how they work together as a trinity and how we might emulate them in AGI.

Language and consciousness disentangled

Say something out loud or, if you are shy, think a (verbal) thought; either will do. Pay attention, please, to each word as it comes out of your mouth or appears in your mind. Ask yourself where each word came from. Start with the first word. Did you have a prior discussion with yourself about what you were going to say or think? If so, from where did the first word in that prior discussion come? Even if you did have a prior discussion with yourself, the only way you could pick each word in advance would be if you committed a series of words to short-term memory. Then, saying those words would just be reciting words from memory, which takes us no further forward.

We talk about picking our words carefully. In light of the introspection suggested in the previous paragraph, what does that phrase mean? For a start, it refers only to when we are speaking out loud. One does not need to pick words carefully when speaking silently to oneself. Nobody else can hear, which is one of the beauties of our inner speech. At most, then, to pick our words carefully means that we speak slowly. Consequently, this gives us the time for the gist of what we wish to say to come to mind first. Alternatively, we have time to hear the next word silently in our heads before uttering it out loud. If we do not like what we hear in our heads, then we stop and wait for an alternative argument or an alternative word to appear in our heads.

This is the nub of it. We do not consciously choose our words, even if we are able to stop them coming out of our mouths when we speak slowly and are on our guards. Rather, the words choose us. They appear, as if from nowhere, in our minds. Sometimes, they are uttered as speech. Most of the time, they are part of the almost incessant stream of words running through our minds whilst awake (and not meditating). Put another way, all these words are lying dormant in our brains. Then something in our brain stimulates a meaningful sequence of them to appear to us consciously. We have a thought, which appears to our conscious mind in the form of words. As noted, this conscious appearance might be part of our continuing inner dialogue or come out of our mouths as speech. If we do not know a word, then it will not appear. Once we have learnt a word, then it might appear if, for example, the flow of a sentence causes it to do so. There is no sense in which we consciously decide what the next word coming into our heads will be.[1] Rather, as David Oakley and Peter Halligan, psychology professors, observe, "I don't know what I am going to say or write next—it simply appears as a thought or verbalization."[2]

This idea does not just extend to words. It also embraces groups of words, termed chunks, that have already been assembled in our minds. Chunks might consist of pithy phrases such as 'the early bird catches the worm'. Chunks even extend as far as concepts and beliefs that we already hold in our heads such as strong views on capital punishment or Brexit. We are all familiar with the admonishment in front of certain people: 'don't say that, you'll just send him off on one.' When he does go off on one, he is merely giving utterance to a pre-packaged series of sentences encompassing the belief in question. The words might vary slightly but the overall meaning is the same.

This account of the nature of language is consistent with the cue-addressable nature of our memories. To recapitulate, memories in a human brain are held in purpose-built neuronal networks. These networks can be triggered or cued by different aspects of the same object or person. Our memory of a long-lost friend might be triggered by their laugh or the set of their jawbone or a particular figure of speech or context. The appearance of words in our minds or our speech is triggered in the same way. A stimulus, often the previous word in the unfolding sentence that we are saying, triggers the appearance of the next word. Consciousness is not directly involved in the generation of the word. Instead, we become conscious of its sudden appearance.

The neuroscientist Christof Koch tells of a man whose brain was partially anaesthetised as part of a medical check-up.[3] This gave rise to his temporary inability to speak or understand language, a condition known as aphasia. He could find no words, out loud or indeed in his head. He saw a tennis racket, knowing perfectly well what it was, but was unable to put a name to it. He remained fully conscious but his brain would not add any words to what he was experiencing. The words just did not appear to him.

This is all really rather surprising, so let us restate it clearly. Language and consciousness are different and separate things. Neither one produces the other. Their main connection to each other is that we become conscious of our language when that language appears in our (conscious) minds. The other important connection between them works in the other direction, from consciousness to language. Specifically, it lies in the role that consciousness can play in promoting thought, particularly new thought, through the interface of language. This role is strictly indirect in nature. We shall return to this theme later in the chapter.

Language and thought disentangled

We considered thought, or cognition, in detail in chapter 7. In summary, there is everyday language and then there is the language of thought, often termed mentalese. As Ray Jackendoff observes, "Thought is a mental function completely separate from language and can go on in the absence of language."[4] Mentalese is a processing language analogous to the operating system code used by a computer. We still do not understand how mentalese works at the neuronal level. But there are theories that purport to explain it conceptually. These include the computational theory of mind and the probabilistic language of thought.

If all the hard work is done by mentalese, then what does everyday language bring to the party? It enables you to communicate in a highly sophisticated fashion with other people. This is a massive advantage that humans have over all other types of animal. Language is an extraordinarily rich tool with which to harvest the fruits of other people's thinking. Broadly, we cannot access those thoughts in any other way. They may contain the answers to superficially unrelated questions of huge importance to people's well-being. Einstein's theory of relativity enabled the Global Positioning System to achieve remarkably high levels of accuracy. That is why your aeroplane lands on the runway as opposed to in a field several miles away. Einstein needed to be able to write down his theory in a way that others could understand. Language, including the language of mathematics, was the only tool with which to do that.

As a result, language does not directly add anything to thought. Indirectly, however, it adds enormously since it is the medium by which other people's thoughts can enter your own thinking. It is also the means by which your own thought process can talk, so to speak, to itself. Language is, among other things, a massive bootstrapping system enabling human thought to climb iteratively to the sophistication we see around us, relative to the rest of nature.

Thought and non-linguistic communication, as exhibited by many animals, are also distinct from each other. Since we do not yet understand the communications system of any non-human animal, it is impossible to determine how rich those communications systems are relative to that of human language. It is also, therefore, impossible to determine whether non-linguistic communication can bootstrap non-human animal thought in any significant way.

If language and thought are distinct, it raises the question how they interact with each other. How does the brain put thought into words and words back into thought? Put another way, the language of thought, mentalese, needs to be translated into words and back again. As Steven Pinker puts it,[5] "Knowing a language, then, is knowing how to translate mentalese into strings of words and vice versa."

Figure 10.1
The language areas of the brain

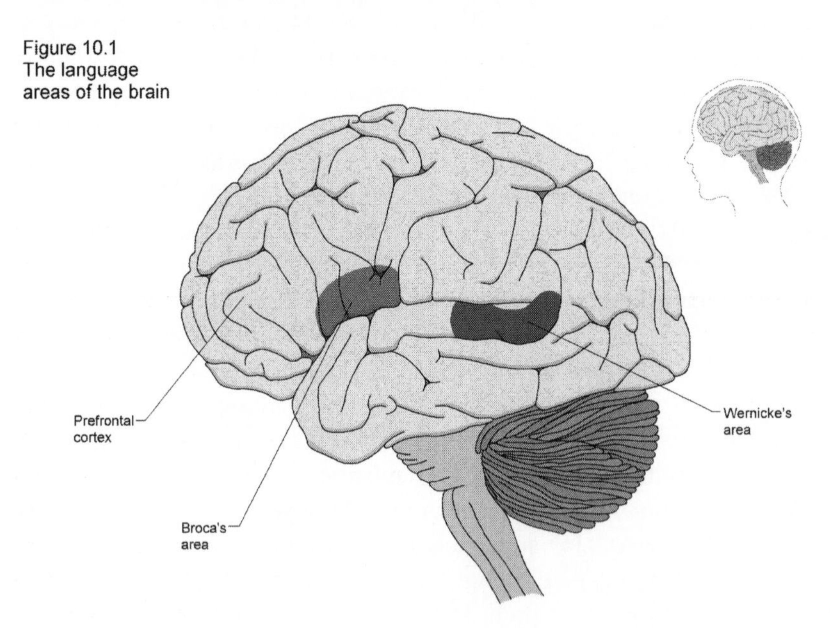

This is the job of brain areas such as Wernicke's and Broca's (see figure 10.1). These two areas of the brain are largely responsible for the comprehension and production of language. Wernicke's area accounts for the majority of language comprehension whilst Broca's area is oriented towards production. This is not to say that Wernicke's and Broca's areas do not perform other roles in the brain too, but language appears to be primarily located in these areas. Other areas of the brain are also involved in aspects of language. For example, the motor cortices of the brain control the muscles activating the tongue, the throat and the jaw and mouth so that you can speak.

We now need to circle back to consciousness and examine its relationship to thought.

Thought and consciousness disentangled

One of the most important conclusions set out in chapter 7 on cognition was that "thought per se is never conscious". The obvious retort to this is, how can we ever know what we mean when we say something if thought is not conscious? The answer is that we are never conscious of the meaning of words; we are only conscious of whether or not they are meaningful.

This is really quite a strange idea at first glance. Jackendoff explains that a conscious thought is made up of three parts.[6] Part one, which he calls "pronunciation", is our consciousness of hearing or perceiving words. Part two is our consciousness of the feeling that those same words mean something. Conversely, we are conscious that a word like 'itytbjyff' is meaningless. Part three is the meaning itself, which is hidden from our conscious brain. The right word, i.e., with the right meaning, generally just tends to appear at the right time in conversation. If it did not, then we would probably be speaking gibberish.

It might be objected that, if asked for the meaning of an everyday word, you could easily supply it. On that basis, the meaning is not hidden from our conscious brain. But, consider: as you speak the meaning of a word in words, each of those words in your definition once again just appears. It feels as if you are conscious of the meaning, but you are not really. All you are conscious of is a "feeling of meaningfulness",[6] a feeling that the word has a meaning rather than being gibberish.

Another objection might be that, if meanings are not conscious, then they are private and confidential to each person's unconscious mind. So, how do we learn the meanings of new words? Surely those meanings are locked up in other people's unconscious minds and, therefore, inaccessible. Strange though it may seem, however, the meaning of a new word goes from the speaker's unconscious mind to your own without ever revealing itself to the conscious mind of either of you. The interface or vehicle for this sharing of meaning is language. The speaker is conscious of explaining the meaning of the word in question and you are conscious of hearing the explanation. You are also conscious of the meaningfulness of the explanation. But the meaning itself slips, like a thief in the night, from one location to another, undetected by either observer. All we see is its shadow in the form of pronunciation and meaningfulness.

Whether or not your unconscious mind accurately captures the meaning depends on the adequacy of the explanation and on your own attentiveness. Sometimes it does not matter if you do not wholly grasp the meaning of something. I only understand the bare bones of Einstein's general theory of relativity, and maybe only one or two of those bones at that. But I can refer to it in general conversation, for example, and that is enough for me. It is much more important that the designer of the navigation system in the aeroplane taking me on holiday understands it. I want to know that his unconscious mind has accurately captured the meaning.

In summary, thought and consciousness are different things. Most of your thoughts enter your consciousness through the medium of language. But it does not have to be that way. I have a serious bridge-playing friend who thinks about many of his moves with no evidence, to him at least, that those thoughts emerge in his mind in the form of language. It is as if he sees rather than verbalises where the cards lie and which is the right card to play when it is his turn. Similarly, the ravens using tools to secure food would, if we grant them consciousness, be aware of a thought process. Language, though, would not have been the interface in that case. Perhaps the interface would have been mental imagery instead of words.

The holy trinity reprised

We have established that language, thought and consciousness are different things. We now need to consider how they work together. With that in hand, we will be better prepared to consider how best to emulate them in AGI. The three of them seem to stand next door to each other in a line (see figure 10.2). Language is in the middle, acting as the interface between thought and consciousness. We can learn more about them and their interaction by considering in which order they evolved.

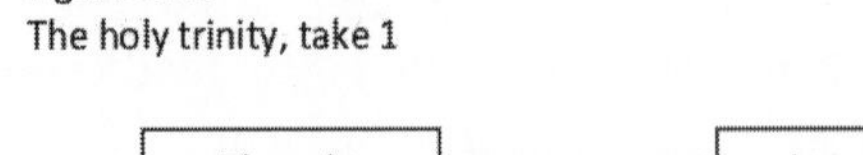
Figure 10.2
The holy trinity, take 1

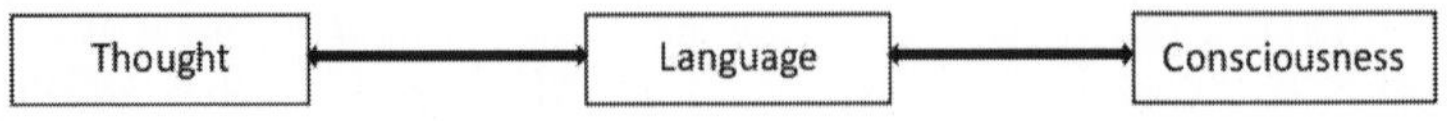

At the risk of getting ahead of myself, I will start by proposing that thought evolved considerably ahead of both language and consciousness. Taking the chronology of thought and language first, all animals engage in greater or lesser degrees of thought of a Type 1 nature. Some animals appear to engage in thought of a Type 2 nature. Yet none of them possesses language. Thus, from an evolutionary perspective, thought must have preceded language.

The more controversial proposal is that thought preceded consciousness too. For a start, there is as yet little agreement on exactly what consciousness is. So, the timing of its emergence must by definition be a matter for debate. The next chapter will consider the nature of consciousness and describe some competing theories. At one end of the spectrum, there are theories that hold that consciousness has evolved relatively recently. At the other end, panpsychism, for example, holds that consciousness is present in most things in the universe, right down to individual atoms. On that basis, consciousness has been around since the dawn of time and must precede thought. It is not possible to prove that panpsychism is wrong any more than it is possible to prove any other account of consciousness is wrong. But I believe that it is more plausible to give an account of consciousness that makes evolutionary sense rather than attributing consciousness to all the atoms in the universe. As will become evident in part 4, I believe that consciousness evolved to enable efficient decision-making in the face of having too many choices. On that basis, thought preceded consciousness.

The remaining question is whether consciousness came before or after language. Let us take a closer look at language.

We do not appreciate the fantastically complicated nature of language when we are learning our mother tongue as a child. Words are learned or memorised with great ease and the minimum of practice. Even syntax, how words are put together in sentences, comes naturally. It is only when we try to learn a new language later on in life that we begin to see how challenging language really is.

Language is composed of a number of features. These include semantics (the study of meaning in language such as the meanings of words), morphology (how words are formed; for example, *smaller* is composed of *small* and *er*), syntax, grammar (syntax plus morphology), phonology (the sounds making up speech), phonetics (how those sounds are physically generated), pragmatics (how context affects production and comprehension of speech) and intonation (how words are given extra meaning as a result of emphasis, volume and so on).

Language entails considerable processing in the brain. It operates at high speed, far faster indeed than consciousness could enable. So, consciousness did, and does, not give rise to language. The question remains, though, which came first? And, what role, if any, does consciousness play in language? For that matter, what role does consciousness play in thought?

Daniel Everett, a professor of cognitive science, notes that "Helen Keller learned to speak only after the age of seven yet in her books she talks about being conscious long before she learned any language."[7] Helen was in the dreadful position of being born deaf, mute and blind. Everett believes that consciousness preceded language. Moreover, he argues that language first emerged some two million years ago in *Homo erectus*.[8] Many scholars believe that consciousness preceded language. This is because many scholars believe that a number of animals have consciousness. Conversely, scholars generally do not believe that non-human animals have language to a human level of sophistication.

Indeed, few would attribute the word 'language' to non-human animal communications at all. This position is coming under some pressure as we learn more about animal communications. The gorilla Koko knew more than 1,000 signs in "Gorilla Sign Language" and could understand more than 2,000 human words.[9] Her companion, Michael, knew about 600 signs. The mere possession of hundreds of different signs and words by two highly trained gorillas, however, does not make standard gorilla communication a language. Language also comprises morphology and grammar, for example, and there is no evidence that gorillas employ morphology and grammar. But that does not mean that animals cannot use their communication systems as interfaces between consciousness and thought. Chimpanzees possess a wide range of calls. We simply do not know whether these are exclusively for communication with other chimpanzees or if they have an internal speech in calls rather than words.

A minority of scholars, I believe, put language before consciousness. The late Julian Jaynes, a psychologist, proposed[10] that the voices of the gods so often heard by the protagonists in Homer's *Iliad* constituted the predecessor to consciousness in what he called the "bicameral mind".

The late Gerald M. Edelman, a Nobel Prize-winning biochemist who later turned his attention to the subject of consciousness, argued that consciousness emerged in stages. It thereby straddled the appearance of language.[11] He proposed that primary consciousness appeared first in a number of animals. The emergence of language in humans led to "higher-order consciousness".

I am proposing that consciousness evolved to lend weight both to the social emotions and to Type 2 processing. I will explain how in part 4. In the meantime, the evidence seems to lean somewhat, rather than fall heavily, in favour of the emergence of consciousness preceding fully-fledged human language. In other words, our holy trinity did not come pre-packaged. First up was thought, as discussed. Then came consciousness and, after that, human language. Consciousness needs an interface in order to optimise its work with thought. Language is that interface. I suggest that an animal with consciousness but without a sophisticated language capability might instead use mental imagery and internalised calls as the interface between consciousness and their thoughts.

A fully-fledged language, however, makes a huge difference to the ability of consciousness to perform its role. You can tell yourself to concentrate and you can keep telling yourself until the task, whatever it is, is done. Try doing that without language. You can tell yourself to calm down and see the other side of the argument. Try doing that without language. Our language abilities have given a strength to our consciousness unmatched in the rest of the animal kingdom. Ironically, it is only owing to language that we can share our thinking about the nature of consciousness. In turn, consciousness, through the medium of language, hones and enriches our thought by promoting our Type 2 processing. It turns out that thought, language and consciousness do not so much stand in a line as stand in a virtuous circle (see figure 10.3). In short, we can control ourselves better than any other animal and thereby unleash a greater emotional and intellectual power than all other living things.

Figure 10.3
The holy trinity, take 2

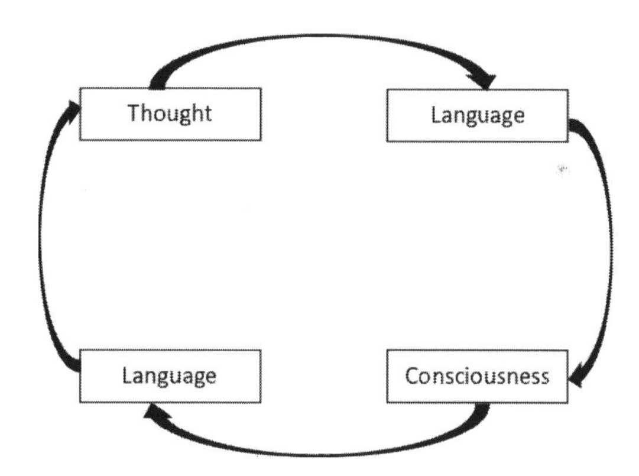

AI's take on thought, language and consciousness

We have already encountered Meta's negotiating chatbots in chapter 3. When negotiating amongst themselves, the chatbots quickly developed a new language incomprehensible to the researchers working on the project. This all sounds very impressive. At a similarly impressive level, the latest natural language processing (NLP) algorithms such as Google's *Meena*[12] and OpenAI's GPT-3 have come close to matching human speech for sense.

But, just as an aeroplane flies but cannot quite match the agility of a sparrowhawk hunting a pigeon, so does NLP successfully generate language but falls slightly short of the flexibility and understanding of human language as spoken by humans.

Humans possess an extraordinary depth and breadth of thought, comprising Type 1 and Type 2 processing. This underpins their facility with language (see figure 10.4a). The problem for AI, or NLP, lies in the relative simplicity of the networks that underpin the production and comprehension of language compared to those in the human brain. NLP does not think in the sophisticated way that a human thinks. NLP networks are one-dimensional and sparsely connected, compared to the massive connectivity in the human brain. NLP typically skips thought in the sense of symbol manipulation altogether. It is a statistical process based on the probability of one word following another. It does not consider causality and counterfactuals; it does not ask, 'what if … instead?' It is not able to make connections between hitherto unconnected objects or events or problems such as the human brain can with transfer learning. It does not possess cue-addressable memory whereby a feature of one object or event can trigger the memory of something else. So, it does not reliably perceive context or exercise common sense. In short, NLP does not interface with a deep and enriched thought process (see figure 10.4b).

Figure 10.4a
How humans use language

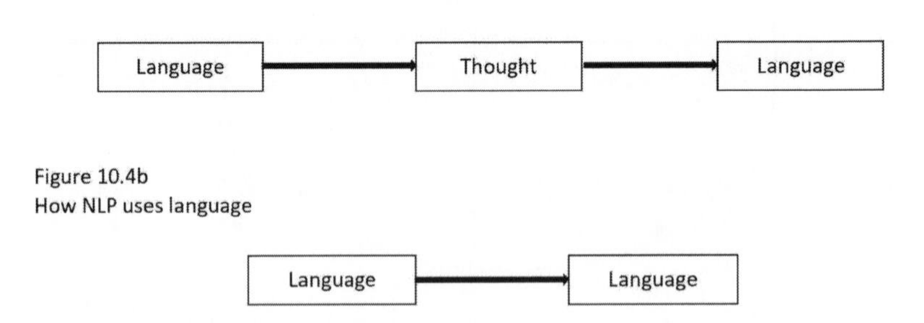

Figure 10.4b
How NLP uses language

As a result, NLP, for the time being at any rate, would be unable to assist the woman in the transcript below in the way that the human at the end of the emergency telephone line did.[13]

Operator: 911. Operator 901. Where's the emergency?
Caller: 127 Glenmere.
Operator: OK, what's going on there?
Caller: I'd like to order a pizza for delivery.
Operator: ma'am, you've reached 911. This is an emergency line.
Caller: yeah, large with half pepperoni, half mushroom.

> Operator: ummm, you know you've called 911. This is an emergency line.
> Caller: Do you know how long it'll be?
> Operator: OK, ma'am, is everything OK over there? Do you have an emergency or not?
> Caller: yes.
> Operator: and you're unable to talk because ...?
> Caller: right, right.
> Operator: OK, is there someone in the room with you? Just say, yes or no.
> Caller: yes.

Encouragingly, a considerable amount of work is taking place in AI circles to rectify the shortcomings of NLP described here. By way of example, a group of researchers has recently developed a successful language generation model called COMET, standing for COMmonsEnse Transformers.[14,15] COMET combines NLP with the symbolic approach to AI.

NLP is not in fact held back by being unaware or not conscious of the meanings of words. We are not conscious of the meanings of words. So, there is no reason for NLP to be. Conscious NLP will not produce better language. In this sense, consciousness in humans and consciousness in AI will achieve subtly different goals. One of the reasons that humans possess consciousness is to promote Type 2 processing so as to be able to engage in symbolic manipulation. There is no reason to conclude that NLP requires consciousness to do the same.

We need to develop conscious AI for another reason, as described at the end of chapter 1. There, I noted that, precisely unless we build ASI with consciousness, ASI will not see the world through our eyes. It will not feel what we feel. Ultimately, it will be unable to empathise with us. That, I suggested, is a wholly unwelcome outcome. Now is the time to look at consciousness more closely and see how it works in detail. Only then can we address how best to instantiate it in AI.

Chapter 11

Selected Theories of Consciousness

A way to consider theories of consciousness

A much-loved debate amongst philosophers over the ages concerns the mind–body problem. On the one hand, we each possess a very real, physical body, of which the brain is a part. On the other hand, we have the intangible, conscious mind, which is much harder to pin down. As David Chalmers observed:[1] "The really hard problem of consciousness is the problem of *experience*. When we think and perceive, there is a whir of information processing, but there is also a subjective aspect. ... This subjective aspect is experience."

The distinction being made in the quotation is between the physical realm and the mental realm, that is, between body and brain on the one hand and mind on the other hand. Body and mind each have states. A physical state, i.e., as found in a body, might be being dehydrated. Your body might well become dehydrated if, for example, you have been walking for a long time in a hot climate. A mental state, i.e., as found in a mind, might be feeling elated or hearing your inner voice tell you to get out of bed in the morning. Try pinning that feeling down; try locating it in a particular place. You cannot. It seems to be in the ether. That is subjective experience.

We can illustrate this distinction another way. In his widely quoted paper,[2] "What is it like to be a bat?", Thomas Nagel asked the question, 'is it like something to be a bat?' There is an element of controversy here in that it is by no means agreed that bats possess consciousness. So, let us stick with, I hope, a less controversial question such as, 'What is it like to be the Prime Minister or the President?' Such a question gives rise to a raft of possible answers. The answers probably all make sense, even if some are not sensible. One might answer, it must feel like running an ultra-long-distance race or it must be like spending one's whole time arranging pieces on a chessboard. But, ask the question, instead, 'What is it like to be a refrigerator?', and there is no answer. It is not like anything to be a refrigerator. A refrigerator just is.

So, there is the physical realm of the refrigerator and your brain. Then, in addition to your brain, but not in addition to the refrigerator, there is the subjective experience of feeling. That could be whether that is the feeling of an emotion or the sensing of, for example, a silent but audible voice in your head telling you to get out of bed in the morning. The CCTV within an alarm system, for example, can see an intruder but, when you or I see an intruder, we also have an experience, or an awareness, of that seeing.

Chalmers contrasts the hard problem of consciousness with a series of "easy problems".[3] These are about explaining, for example, "the ability of a system to access its own internal states" and "the focus of attention", all of which abilities we humans perform quite naturally. A non-conscious machine can do any and all of these things too. A computer can, and frequently does, access, or examine, its own internal states. These might include its current speed of processing, how much random-access memory is being consumed and so on. A robot on a manufacturing line will be focussed on, or attending to, first one spot and then another. What is hard, which is why Chalmers calls it the hard problem, is why, when we humans think or perceive or emote, it feels like something. After all, when a (non-conscious) machine processes information, or in human-speak thinks, the act of processing does not feel like anything for the machine.

This is not just a question posed out of idle curiosity. Setting aside the question whether other animals possess it, we all feel deeply that consciousness lifts us above mere inanimate objects. We cannot wish consciousness away. We will be unable to produce authentic and really helpful artificial human-level intelligence without fully understanding consciousness.

There is an increasing number of theories of consciousness. We will, therefore, look at only a few of them in this chapter. Before turning to that task, it is worth setting down some tests for what makes a good theory of consciousness. Whilst none of these tests is absolute in nature, they constitute hurdles that any theory of consciousness ought to be able to surmount (see figure 11.1). If a theory cannot do so, then it probably needs extra scrutiny, if indeed it survives at all. This is perhaps one way of assessing the quality of a theory of consciousness.

Figure 11.1
The four tests for a theory of consciousness

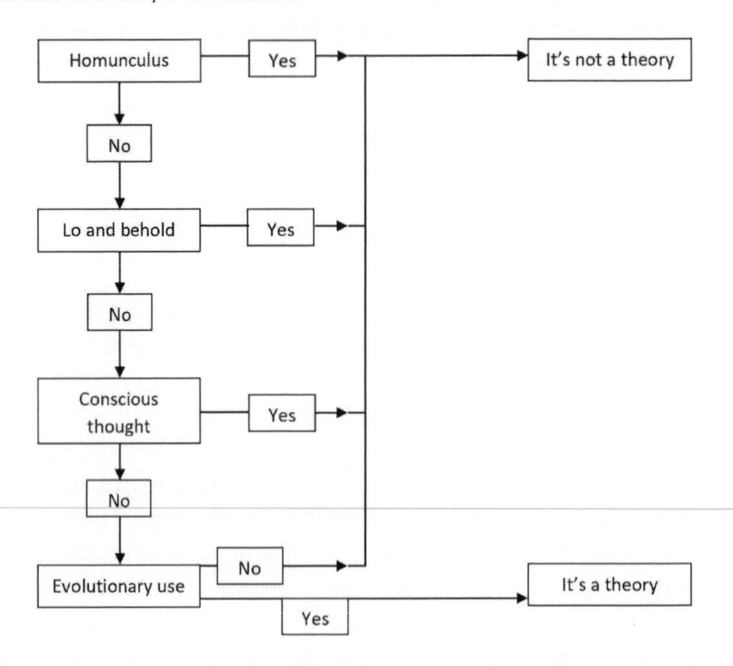

Let us start with the homunculus test. In other words, is the theory in question really just a proposal that there is a homunculus sitting inside our brains? For British readers of a certain age, the idea of the homunculus was rather well captured by *The Beezer* comic strip, the *Numskulls*.[4] In the comic strip, the human, called Our Boy, has six numskulls sitting inside his head. Five of the numskulls are responsible for the senses of sight, hearing, taste and smell. The sixth numskull, Brainy, is responsible for operating Our Boy's brain.

The problem with the homunculus theory, as described in chapter 8, is that it raises the question, how does the homunculus do its viewing and deciding? Or how does Brainy decide how Our Boy should behave? The answer can only be that inside Brainy's head there is in turn another Brainy doing the deciding, and so on. In other words, the proposal ultimately regresses into absurdity. So, the test in this case would be that, if upon analysis the theory in question contains a homunculus, then the theory is probably absurd.

One way out of the dilemma of an ever-increasing number of homunculi is to identify the homunculus as a soul or some such immaterial entity. This is the approach taken by many religions. I do not consider it further here.

The second test is labelled the lo and behold test. There are two alternative dimensions to this test depending upon the nature of the theory under inspection. The first dimension concerns any theory that as good as asserts that consciousness just is. The second dimension concerns any theory that says that consciousness comes into being as a result of the combination of certain constituent elements.

The first dimension to the lo and behold test is addressed by the idea of falsifiability. This derives from the work of Karl Popper, one of the greatest of philosophers of science. In essence, Popper held that a theory could only be counted as scientific if it were possible to imagine how it might be falsified. Take the idea that Satan lives in the kingdom of hell. This cannot be held to be a scientific theory since it is not falsifiable. How on Earth is one ever going to be able to locate hell in the first place, let alone track down Satan? To return to our black and white swans from chapter 1, the theory that all swans are white was a valid scientific theory. One could imagine finding a swan of another colour and, therefore, the theory was falsifiable. So, in this dimension, the lo and behold test flushes out theories that look as if they cannot be falsified and are, therefore, not scientific.

The second dimension to the lo and behold test is exemplified by the idea of emergence. Many theories of consciousness ultimately come down to the idea that consciousness is an emergent phenomenon. An emergent phenomenon arises as the result of the coming together of smaller, interacting constituent elements. Indeed, it is often the case that one cannot easily predict emergent phenomena by studying their constituent elements. If one had never seen or heard of a snowflake, then extensive analysis of hydrogen and oxygen atoms would probably not give rise in one's mind to the possible formation of snowflakes. But, knowing about hydrogen and oxygen and knowing about snowflakes, the scientific community has been able to explain how the former can transition to the latter. Chalmers identifies two types of emergence, weak and strong.

Weak emergence exists when we can reduce the phenomenon to the properties of its constituent elements. After the good work done by Johannes Kepler,[5] a famous seventeenth-century astronomer, and others, we can reduce the emergent snowflake to its constituent elements. The snowflake only appears to be irreducible to its elements at first sight. It is in fact weakly emergent.

Strong emergence exists when a phenomenon arises out of its constituent elements but truths concerning that phenomenon are not deducible even in principle from features of those constituent elements.[6] Put another way, it is not possible to reduce the strongly emergent phenomenon to the properties of the constituent elements of the phenomenon. Strongly emergent phenomena then are almost magical. Scientific knowledge these days is such that it is hard to think of anything complex that cannot be explained by reference to its underlying parts, provided that it has underlying parts. Chalmers believes that there is exactly one clear case of a strongly emergent phenomenon, and that is the phenomenon of consciousness.

We shall come back to Chalmers' theory of consciousness shortly. The point, however, is that if a theory of consciousness is emergent in nature then we need to be on our guard. A theory might be based on weak emergence but fail to reduce its description of consciousness to the properties of its constituent elements. It might, in other words, leave consciousness at an immaterial level; hence, lo and behold. Alternatively, a theory might be based on strong emergence but fail to explain adequately how it reaches those lofty heights. The difficulty with calling consciousness a strongly emergent phenomenon is not that it might not be true—it might be. It is that, on its own and without explanation, it is no more than a *deus ex machina* or lo and behold moment. It is an assertion, not a theory. For that reason, it has drawbacks. A theory should at least try to bridge the gap between the relevant constituent elements and consciousness. We can then subject the theory to appropriate testing.

The various theories of consciousness described below will generally illustrate that it is hard to avoid the lo and behold moment. Either they are emergent theories or they in essence assert that consciousness was more or less there all along.

The third test derives from the extensive role of the unconscious mind in the brain's workings. As we saw in chapter 10, thought is unconscious. I cited Ray Jackendoff's work relatively extensively on this point. He himself notes that there is a large number of thinkers who would hold a different view. He mentions, among others, Bernard Baars, Antonio Damasio, John Eccles, Christof Koch, Gerald Edelman, Douglas Hofstadter and Giulio Tononi.[7] This is admittedly an impressive list, but merely being impressive is not enough. Thus, our third test is that a theory of consciousness must exclude terms such as conscious thought in favour of terms such as consciousness of thought.

The fourth test is the evolutionary use test. As John Searle, a philosopher with many years' standing in the field, puts it:[8] "The processes of conscious rationality are such an important part of our lives, and above all such a biologically expensive part of our lives, that it would be unlike anything we know in evolution if a phenotype of this magnitude played no functional role at all in the life and survival of the organism." In other words, consciousness is such a significant part of our lives that it would not have persisted from an evolutionary perspective, unless it were achieving some useful goal.

Of course, this presupposes that consciousness is a product of our genetic material, albeit expressed in a particular way and perhaps shaped by our upbringing. More broadly, it presupposes that consciousness is only possessed by organisms, that is, living entities. This reflects Searle's so-called "neurobiological approach to consciousness[, which] eschews the possibility that consciousness could be embodied in other kinds of inanimate material".[9] It also presupposes that consciousness is causal in nature. These premises are at odds with some of the theories of consciousness summarised below, which would make it an unfair test. It might be better termed a caution than a test. So, with that qualification, my fourth test would be that a respectable theory of consciousness needs to include a satisfactory explanation of what consciousness does for its possessor. It needs to answer the question, 'What is consciousness for?' If it does not, then inspect it more closely.

David Chalmers

In his seminal book *The Character of Consciousness*,[1] Chalmers segments theories of consciousness into six classes. These divide into two master-classes, *reductive* and *nonreductive*. A reductive account of consciousness explains it, and particularly its subjective or phenomenal aspect, purely by reference to physical principles. A reductive account of a snowflake would explain it by reference to its constituent elements, i.e., hydrogen and oxygen, and how they combined to form the snowflake. A non-reductive account of consciousness, on the other hand, preserves or asserts its phenomenal or subjective aspect as part of the explanation, in other words over and above its constituent elements.

Put another way, consciousness is one of two things, physical or non-physical. In the first case, consciousness is purely physical but appears non-physical. This is a reductive account. In the second case, consciousness is, as a matter of fact rather than merely appearance, non-physical. That is to say, consciousness is distinct from physical things. This is a non-reductive account. In other words, such an account does not reduce consciousness down to the purely physical explanation. This distinction lies at the heart of all theories of consciousness.

Thus, in the final analysis, a reductive explanation of consciousness denies the existence of any immaterial aspect to consciousness. Chalmers points to three types of reductive explanation. "Type-A materialists" acknowledge no difference between, for example, seeing and the feeling of seeing or being angry and the feeling of being angry. They place function over form. The hurdle for type-A materialism is the lingering feeling that there remains the feature of feeling. "Type-B materialists" acknowledge the existence of feeling but conclude there is a way of explaining it in physical terms. The difficulty with this approach is that it is one short step to being a nonreductive account of consciousness, as we shall see when we come on to "type-D dualism". "Type-C materialists" perceive a real role for consciousness. Yet they also claim that the true account of consciousness must be reductive; they have just not yet figured out what that account is.

Nonreductive accounts, however, embrace the immaterial feature of consciousness. "Type-D dualists" see two interacting realms: the physical and the phenomenal. Each can cause change in the other. In today's world, based as it is on a certain picture of physics, such an account might appear improbable. Yet, a type-D dualist would argue that the phenomenal realm is simply a fifth fundamental force in addition to the weak and strong nuclear forces, electromagnetism and gravity. "Type-E dualists" go one step further. The phenomenal realm is so different that it does not interact with the physical realm at all. It just comes along for the ride, so to speak. This is called "epiphenomenalism". Finally, "type-F monists" assert that the phenomenal realm is intrinsic to physical reality, that it in fact applies at the sub-atomic particle level. The type-F approach looks like type-D dualism but Chalmers distinguishes them by locating type-F monism below "the micro-physical level" and type-D dualism above it.[10]

Chalmers himself leans towards a nonreductive account of consciousness.[11] All three nonreductive accounts of consciousness find it hard to address the lo and behold test because it is difficult to show that such theories are falsifiable. Type-E dualists, however, find it especially hard since they provide no explanation at all. Nonreductive accounts of consciousness also appear to sidestep the evolutionary test altogether. In addition, type-D dualism falls foul of the homunculus test.

Chalmers' taxonomy of consciousness is so useful that I will label each of the following theories of consciousness according to it in order to assist comprehension. We will go back and forth from one side of the reductive–nonreductive divide to the other in order to help highlight the differences between the theories. In that vein, let us turn from Chalmers' own nonreductive belief to an apparent type-A materialist, Daniel Dennett.

The rest of this chapter is more technical and can be skipped if the reader prefers.

Daniel Dennett

Dennett has written extensively on consciousness, with numerous books and papers to his credit. Perhaps his most comprehensive account is that contained in *Consciousness Explained*.[12]

The first half of the book is given over to the demolishing of the "Cartesian Theatre"[13] and the installation of the "multiple drafts model".[14] The Cartesian Theatre, inspired by Descartes, is a variation on the homunculus approach; it is Type-D dualism. It claims that there is a small portion of the brain that houses phenomenal or subjective processing. Indeed, Descartes identified that portion of the brain as the pineal gland, which is located in the middle of the brain. In fact, the pineal gland produces melatonin and thereby regulates sleep patterns. Dennett's multiple drafts model replaces the Cartesian Theatre altogether. It proposes that the brain's neuronal networks generate numerous potential perceptions, thoughts and actions, i.e., the various drafts. In a winner-takes-all way, just some of those wind up in consciousness.

Dennett then tackles how consciousness arose in the first place.[15] He suggests that one possible route to consciousness was the internalisation of "auto-stimulation". He gives as an example of auto-stimulation an early (pre-conscious) hominid asking another hominid for help on a certain question. As it happens, the second hominid has gone off somewhere else and the first hominid finds itself answering its own question. Over time, the first hominid discovered that it did not need to waste resources speaking out loud to itself. Rather, it internalised the process. Consciousness then is a matter of acknowledging or reporting the winning drafts out of the numerous available drafts. These are the potential perceptions, thoughts and actions generated by the brain's neuronal networks. We are, however, still left with Chalmers' hard problem: why should this internalised, reporting self feel like anything?

Without further development, Dennett's theory could feel as if it were lacking something fundamental. With a nod to Nagel's bat, he explains that the reason that consciousness feels like something is because that is what happens when the brain perceives itself.[16] This is, however, a "user illusion".[17] In other words, the brain has tricked itself into the perception of an internalised self. This, we call consciousness. Such trickery by the brain is not remotely unusual. When the brain receives the input from our eyes when they see a table, the brain does not deliver up a vision of countless millions of molecules of wood matter, let alone the underlying atoms of carbon, hydrogen and oxygen. That would be bamboozling. Instead, you see a table. It does not matter that the internalised self is an illusion because it enables "self-control".[18]

Dennett could be seen as a type-A materialist, which from an intuitive perspective is not an easy position to hold. In thinking about it, there are various technical arguments that purport to strike down type-A materialism.[19] Fundamentally, to be a type-A materialist, one has to distrust, if not discard, one's subjective experience. Nonetheless, Dennett's portrayal of consciousness as a user illusion might offer an elegant way round this problem.

Dennett's account passes the evolutionary use test. In so doing, however, it appears to fall foul of the conscious thought test given the proposed role of consciousness in self-control. Dennett's identification of self-control as being important is interesting but needs to be explained further, as we shall see in chapter 17. Dennett admits that his account could be drifting into homunculus territory[20] while insisting that it is not. The idea of a user illusion is also interesting and we shall also return to this idea. Let us now switch to type-D dualism.

Benjamin Libet

One of the biggest problems with consciousness as a topic is that the amount of hard neuroscientific research into it is small. It is difficult to research something that, so far, we have been wholly unable to locate physically. The late Benjamin Libet, a professor of physiology and a pioneer in the arena of consciousness in humans, has actually tested the feature of consciousness head-on. His core finding is so startling that it bears quoting from his book, *Mind Time*, in full.[21]

> We were able to examine this issue experimentally. What we found, in short, was that the brain exhibited an initiating process, beginning 550 msec before the freely voluntary act; but the awareness of the conscious will to perform the act appeared only 150–200 msec before the act. The voluntary process is therefore initiated unconsciously, some 400 msec before the subject becomes aware of her will or intention to perform the act.

Moreover, he observes that the pre-planning for the act might well have happened even further in the past than the minus 550 msec referred to above. That would be consistent with the value-based decision-making process described in chapter 8. What he is suggesting, therefore, is that there is no such thing as a conscious decision to act. He proposes instead that there is only an unconscious decision to act, of which we then become conscious, or aware.

Libet also postulates that the interval between the point at which we become aware of the decision to act and the actual act itself is not necessarily left idle. The interval of time is 150–200 msec. He suggests that this is long enough for us to veto the (unconscious) decision to act if we so wish. In other words, we have a "conscious veto".[22] This is contentious, as Libet himself realises. In fact, Libet's interpretation of his findings are almost more startling than the findings themselves. He asks the question, "*Does the Conscious Veto Have a Preceding Unconscious Origin?*" If it did, then our so-called conscious veto would be under the control of the unconscious mind in the same way as our decisions to act. At that point, one could hardly call it a conscious veto. Instead, surprisingly, he cites "conscious will" as being a force in its own right, able to act as a veto.

Libet is in danger of self-contradiction. He makes the point, rightly, that "Vocalizing, speaking, and writing ... are all likely to be initiated unconsciously".[23] In other words, thinking (in this case, verbal) is unconscious. Yet, all of a sudden, "the actual decision to veto",[24] which is just a type of thinking, is apparently conscious. That is hard to understand.

Libet proposes an answer to this challenge. He emerges as a type-D dualist, the first of Chalmers' nonreductive approaches. He suggests that there is a fifth fundamental force, called the "conscious mental field" (CMF).[25] He suggests that the CMF might have causal properties, which would potentially allow the conscious veto to operate. This is the area in which dualism typically faces its biggest challenge: how can something immaterial have any sort of interaction with the physical realm? Libet tries to circumvent the mind–body interaction problem by defining CMF as a fundamental force, like gravity or electromagnetism.

In summary, then, Libet's primary scientific finding is of significant importance, although it should be noted that even this finding has been questioned by some researchers. Libet's theory of consciousness looks as if it scores half marks on test three (conscious thought), with a 50% deduction due to the suggestion of a conscious veto. He probably side-steps test one (homunculus), depending upon how one perceives the CMF. Libet's theory finds test two (lo and behold) difficult. It is hard to see how the idea of a CMF is falsifiable. It is not that one can disprove the existence of a fifth fundamental force. It is rather that, without any supporting evidence, it does not yet feel like a satisfactory explanation.

Edmund Rolls

Rolls' higher-order syntactic thought (HOST) theory of consciousness is part of a family of higher-order theories of consciousness. Such theories hold that a system, e.g., a human, has a conscious thought when that system thinks about that thought.[26] In other words, an unconscious first-order thought becomes conscious when a higher-order thought (another unconscious thought) thinks about the first-order thought. Self-consciousness arises when an unconscious higher-order thought becomes conscious because an unconscious, even higher-order thought thinks about the first higher-order thought. This sounds circular in that it is not far from saying that a thought is conscious when it is conscious. Higher-order thought theories of consciousness are type-B materialist.

Rolls is a computational neuroscientist, as well as being a psychologist. His HOST theory takes the basic higher-order theory of consciousness one step forward. On his account, consciousness evolved to enable the brain to solve a processing problem. What happens when the brain is given the challenge of solving a complex reasoning problem? Rolls takes us back to one of the core tasks of the brain, decision-making, as already discussed in chapter 8. A simple decision might involve a single autoassociation neuronal network (see chapter 5). A bigger challenge might arise when there are two possible solutions that are almost evenly matched. The answer to the challenge is a second network. A second, nearby neuronal network might measure the "confidence"[27] of the decision arising from the first network. Should the confidence be inadequate, the decision might be sent back to the first network for review. From a computational perspective, this is straightforward. Thus, many complex reasoning problems look as if they can be solved with a relatively straightforward addition to the brain's architecture.

The processing problem arises when a "multi-step reasoned plan"[28] has to be conjured up. The brain might swiftly come up with such a plan, especially if it is reliant on Type 1 processing or fast thinking. But there may be holes in the plan. How can the brain figure out which step in the plan might be faulty? Rolls terms this the "credit assignment problem".[29] Building on philosopher David M. Rosenthal's work,[30] Rolls suggests "that by thinking about lower-order thoughts, the higher-order thoughts can discover what may be weak links in the chain of reasoning at the lower-order level, and having detected the weak link, might alter the plan, to see if this gives better success".[31] The syntactic nature of the higher-order thoughts enables them to evaluate each link in the chain of reasoning so as to rectify any weaknesses. The really big step in the theory is that, in order for the higher-order thoughts to evaluate the links in the chain effectively, the first-order thoughts need to become conscious.[32]

Rolls' theory is grounded in the workings of neuronal networks, particularly with respect to decision-making. This is wholly plausible since we have already seen that neurons and neuronal networks are at the heart of how the brain works. So, rooting a theory of consciousness in neuronal networks makes eminent sense. Rolls' theory neatly passes test four (evolutionary use). It might also be seen to pass test three (conscious thought). The higher-order thought, i.e., the one that does the work, is not conscious. Rolls' HOST theory says that, in order to do its work, it merely has to be directed at a conscious, first-order thought. This, however, takes us straight into test two (lo and behold). Why should a first-order thought be conscious; why does it need to feel like anything at all? It does not solve Chalmers' hard problem, as Rolls more or less concedes when he refers to "an element of 'mystery'" in the argument.[33]

Bernard Baars/Stanislas Dehaene

The global workspace theory of consciousness (GWT) was first unveiled by Bernard Baars in 1989.[34] It went on to be elaborated in neuronal terms by Stanislas Dehaene and colleagues in 1998.[35] GWT proposes an identity between the contents of consciousness and information that is broadcast across the brain by way of the global neuronal workspace. Put another way, a perception is or becomes conscious when its representation in neuronal terms is made widely available in the brain.

Many perceptions might become conscious, but only some do. They compete with each other in the unconscious brain to do so. But at any point in time only one succeeds when it triumphs in a winner-takes-all computation.[36] The winner "is selected because of its relevance to our present goals".[37] Having won, that conscious perception is broadcast across the brain so that other, unconscious areas of the brain can work on it. It might be another piece in a puzzle for which the brain is trying to find a solution. It might be another note in a sequence of music that the brain is trying to learn off by heart. Dehaene proposes that the workspace is realised by "a dense [long-distance] network of interconnected brain regions".[38] He has also found that there is a match between the activation of this network and the reports by experimental subjects of conscious perceptions or thoughts.[38,39]

GWT is a functionalist theory of consciousness, a sort of what you see is what you get approach to the subject. As psychology professor Susan Blackmore points out, GWT could be taken in one of two ways.[40] One way of looking at it is to propose that the mere fact of something being broadcast in the brain causes it to become conscious. On this analysis, it would be epiphenomenal. Dehaene, however, rejects the idea that it is epiphenomenal. The other way to look at it is to propose that "the broadcast is all there is".[41] That approach rather takes the wind out of subjective experience's sails. Dehaene goes on to say: "Like the psychologist Bernard Baars, I believe that consciousness reduces to what the workspace does..."[41] On that basis, GWT is a theory of cognition rather than a theory of consciousness. It falls into type-A materialism.

Like HOST, GWT, especially as elaborated by Dehaene, is grounded in neuronal networks. The idea that pertinent information bubbles up into consciousness by way of a winner-takes-all process and then loops back into the unconscious brain is useful. We shall return to it in part 4. It looks, however, as if it fails either or both of tests one and three (homunculus, conscious thought). Dehaene avers that GWT does not rely on a homunculus.[42] But he also suggests that *we* can use the contents of the workspace "in whatever way we please".[43] This rather raises the question of what *we* means. If *we* is not meant to be a homunculus, which would otherwise fail test one, then at the very least that gives the global workspace the power of conscious thought, which fails test three. If conscious thought is also rejected by GWT as a consequence of the work of thought being done unconsciously, then we are back to our conclusion that GWT is just a theory of cognition. There is no need for subjective experience. On that basis, test two (lo and behold) is passed precisely because there is no lo and no behold. It does not solve Chalmers' hard problem. On that basis, GWT does not pass test four (evolutionary use) because it does not even sit that part of the exam.

Philip Goff

Borrowing loosely from Searle,[44] Scylla and Charybdis were sea monsters in ancient Greek mythology between which seafarers had to pass in order to sail between Sicily and mainland Italy. The sea that is philosophy of consciousness has the same monsters. Scylla represents the mind–body interaction problem and threatens to sink dualism. Dualism says that things are made of physical and phenomenal substances, or possess both physical and phenomenal properties, sitting side by side and interacting with each other. Charybdis never lets you forget the constant presence of subjective experience and thereby threatens materialism. Which wily ship of philosophy can sail between them without being wrecked? According to Philip Goff, a philosopher, that ship is panpsychism, a form of type-F monism. This is the belief that pretty much everything, including sub-atomic particles such as electrons, possesses consciousness. Put another way, their properties are simply forms of consciousness. It feels like turtles all the way down.[45]

In his book *Galileo's Error*,[46] Goff argues that panpsychism avoids the mind–body interaction problem of dualism because consciousness is inherently a feature of the brain due to its constituent elements at the minutest level. Panpsychism does not suggest that, separate to the physical, there is some other non-physical stuff. In contrast, panpsychism says that "the physical properties of a particle (mass, spin, charge, etc.) *are themselves* forms of consciousness".[47]

Goff also argues that panpsychism avoids the charge of reductionism levelled at materialist theories of consciousness. Reductive theories of consciousness, especially those in the type-A category, risk dismissing the phenomenon of subjective experience altogether. Panpsychism is non-reductive. He says there is nothing wrong with that. We can only explain gravity in nonreductive terms, for example.

This sounds a little like one of those pro having cake and pro eating cake moments.[48] Goff blames Galileo for having made a big mistake some 400 years ago, which has led to confusion ever since. He says that Galileo located the world of science in a purely quantitative realm. This thereby "confines itself to telling us only what an electron [or anything else] does".[49] He says that an electron must also have "an *intrinsic nature*"[49] This is not about what an electron does, but rather about what it is "in and of itself". In fact, because we normally explain everything in terms of what it does, all our existing explanations of things are ultimately circular. A panpsychist instead attends to the intrinsic nature of things, including consciousness, such that "a feeling is not defined in terms of anything other than itself."[50] What physics has done, inadvertently, is to fail to see that an intrinsic inspection or definition of things would reveal that they all possess consciousness. This includes brains.

Goff acknowledges that the big problem for panpsychism is the so-called "combination problem ...: How do you get from little conscious things, like fundamental particles, to big conscious things, like human brains?"[51] What in between those two ends of the spectrum is conscious: a pair of socks, a thermostat, a rabbit? This remains an unsolved problem. Some panpsychists try to address the challenge by calling upon the idea of intrinsic emergence. Intrinsic emergence is "the emergence of new forms of conscious intrinsic nature".[52] In other words, as the sub-atomic particles of the brain come together, each of them individually conscious in their own right, their very coalescence gives rise to a conscious whole, the conscious brain. This sounds like Chalmers' strong emergence, along with its attendant difficulties.

Since panpsychism lays claim to the entire physical and phenomenal realm, it could be said not to fail test four (evolutionary purpose). If consciousness was there all along, then it did not need to evolve. On the same grounds, it does not so much fail tests one and three (homunculus, conscious thought) as not see them as appropriate tests in the first place. Panpsychism effectively shifts the goalposts. It is in a sense a materialist theory, which might place it into type-B, for example. It is, though, explicitly not reductive, which is why it is in fact a type-F theory.

Panpsychism cannot dodge test two (lo and behold). It is not that one can disprove the existence of ubiquitous consciousness. It is rather that, without any supporting evidence, it does not yet feel like a satisfactory explanation, even if it is an elegant one. It is not obviously falsifiable. Panpsychism involves the modification of, or addition to, modern physics. As Sean Caroll, a research professor of physics, observes, we can ask that "any modification of the [now widely accepted Core Theory of physics] be held to the same standards of rigour and specificity that physics itself is held to."[53]

Antonio Damasio

Damasio is, like Rolls, a neuroscientist. He has treated consciousness as a major subject in its own right and has written extensively about it. At the heart of his account is the evolution of the self. He is a type-B materialist.

Damasio proposes that consciousness evolved in stages, like most features in our brains and bodies. But consciousness is in turn a product of the evolution of the self, which also developed in stages. Stage one was the "protoself".[54] This takes us back to the loose distinction I drew in chapter 5. The simplest organisms merely sense and move. In complex organisms such as *Homo sapiens*, the order of play is sense-process-move, with the emphasis on process. But the brain can only process if it has content and an elementary piece of content is a neuronal network, or a collection of networks, depicting the organism's own body. Put another way, the brain can only process with a view to making good behavioural decisions if it has a picture of, among other things, its own body. This is the protoself.

But, and this is the critical step in the theory, it is not enough that the brain generates the protoself. For the protoself to be effective as part of the neuronal machinery driving behaviour, it must be "*felt*" too.[54] Damasio terms this felt body image "primordial feelings of the body". He claims that these feelings are "spontaneously present"[54] and attributes their emergence to the "complex interconnectivity of [these] brain-stem nuclei", i.e., where body interacts with brain.[55] Given that he is talking about the first step in an evolutionary account of consciousness, "numerous living species" turn out to possess this particular "manifestation of consciousness".[56]

After the protoself comes the "core self".[57] The core self underpins core consciousness. From the description above, the protoself comes across as essentially passive. In contrast, the core self is a "protagonist".[58] Damasio proposes that the core self evolves from the protoself as a result of the protoself becoming able to adjust, neuronally, to the sensing of perceived objects. The new, emergent core self is literally changed by what it perceives. The brain has evolved to incorporate a feedback mechanism. The brain does not just know the object perceived; it has a "feeling of knowing the object".[59] In other words, core self takes the form of two selves, Self-as-object and Self-as-subject.[60] The protoself was the forerunner of Self-as-object. Self-as-subject has emerged in core self as a parallel form of self able to perceive and interact with Self-as-object. Whilst this sounds a little mystical, Damasio is clear that both these selves are in point of fact neuronal networks that interact with each other.[61]

Damasio further proposes that the core self is made manifest in "pulses".[62] In other words, at this level of consciousness, the feeling of knowing the object is not smooth. It is more akin to the individual frames in a clip of film that has been slowed down. One can (just about) imagine "core consciousness" as constituting an unending series of disjointed fragmentary moments. This clearly differs from the smooth joined-up subjective experience to which we are nowadays accustomed. These disjointed moments are Damasio's pulses.

The pinnacle of consciousness, "autobiographical consciousness", i.e., such as humans possess, is achieved on the back of "autobiographical self".[63] Autobiographical self evolves from core self when our memories are able to be presented in the mind in the same way as perceived objects. Additionally, the pulses of core self become coordinated by the brain so that they run smoothly into each other to create the continuous stream of consciousness that most of us take for granted.

Like Rolls, Damasio seeks to embed his theory deeply in the workings of the brain. He goes on to present reasons why consciousness enhanced the fitness of the organism. In the first instance, primordial feelings evolved because states of emotion "actually *ought* to feel like something"[64] in order to drive behaviour. At the other end of the spectrum, autobiographical consciousness enabled "planning and deliberation".[65] On this basis, Damasio absolutely passes the evolutionary use test. But, in so doing, his account fails test three (conscious thought). Test one (homunculus) is successfully hurdled. Questions would be raised around test three (lo and behold): he floats the idea that feelings arise from the interconnectivity of large numbers of neurons in a brain.[66] This is strong emergence with all its problems. This is exactly where type-B materialism shades into type-D dualism. The question I posed at the beginning of the chapter was whether an account of consciousness that relied on emergence was accompanied by a satisfactory explanation. It is not obvious that pointing to the interconnectivity of large numbers of neurons is adequate.

Importantly, Damasio's account emphasises the role of self. He divides self into two highly inter-related constituent pieces, Self-as-subject and Self-as-object. This is an interesting distinction, upon which we will build in chapter 15.

Giulio Tononi

Giulio Tononi is the originator of the integrated information theory of consciousness ('IIT').[67] IIT is based on the idea that it can describe consciousness in a very precise way. As a result, it can then in principle measure consciousness in whatever form it takes. If, in turn, those measurements bear up to empirical scrutiny, then IIT has proved its worth.

IIT's description of consciousness, therefore, is of great importance. It claims to describe it fully with five axioms. Any conscious "experience exists for itself, is structured, is the specific way it is, is one, and is definite".[68] Each axiom gives rise to a postulate that purports to make the axiom tangible, leading to the ability to quantify the experience in question. The first of these axioms ("experience exists for itself") proposes that a conscious experience has an intrinsic nature. There is a difference between IIT's use of the term 'intrinsic' and that of panpsychism. Panpsychism invests intrinsic with a fundamental quality at the sub-atomic particle level. The postulate that flows from IIT's first axiom proposes, however, that experience occurs in a system with "cause–effect power *upon itself*".[69] Tononi *et al.* illustrates the idea thus: "A minimal system consisting of two interconnected neurons satisfies the criterion of intrinsic existence because, through their reciprocal interactions, the system can make a difference to itself."[70]

The postulates also suggest that the amount of consciousness of a system can be measured. A system for these purposes might be a particular combination of neuronal networks (hence the term 'integrated information') giving rise to a conscious experience. Thus, each conscious experience such as reading a word on the page or viewing a landscape will have a certain measure of consciousness. IIT labels this generically as Φ. Φ is always taken as measuring the maximum amount of integrated information in the system. Any neurons that might be connected to the system but do not contribute to the experience in question, i.e., that are not integrated, are excluded.

On the face of it, like panpsychism, IIT neither passes nor fails test four (evolutionary use). It "takes no position on the function of experience as such".[71] It claims, however, to answer the question 'why' in another way. In virtual environments used for testing IIT, there appears to be a correlation between the amount of integrated information in a participating virtual entity and its success at navigating that environment.[71] Moreover, there is also significant correlation between "the brain's capacity for information integration" and the state in which it is. Thus, the brain state when the subject is asleep, for example, shows highly diminished integrated information, consistent with the concomitant loss of consciousness.[72] Like panpsychism, IIT sidesteps tests one and three (homunculus, conscious thought).

IIT is in a sense a materialist theory, which might place it into type-B. Indeed, Christof Koch, who has worked extensively with Tononi, seems to dismiss the importance of "subjective feelings" to consciousness altogether.[73] In his, and Tononi's, view, consciousness is wholly and completely captured by the five axioms. It is not, however, clear that the five axioms completely define consciousness. It is not as easy as that to slip past subjective experience unnoticed. IIT concentrates on correlation rather than causation and thereby seeks to evade Chalmers' hard problem. As we saw in the discussion of the link between smoking and lung cancer in chapter 7, the focus on correlation to the exclusion of causation can lead to erroneous conclusions.

Whilst materialist, IIT is explicitly not reductive, which is why it looks more like a type-F theory. Koch, for example, suggests that atoms may have conscious experience.[74] It fails test two (lo and behold) precisely because it defines itself by reference to its own conclusion. So, it seeks to sidesteps the idea of falsifiability. Put another way, the five axioms may lead to the results discussed above but they presuppose they have accurately defined consciousness in the first place. It is like a magician with two hats, who directs his audience to concentrate on the first hat while pulling a rabbit out of the second.

The theories of consciousness described above illustrate the fact that consciousness is indeed a hard problem. Moreover, this is just a small selection of theories: there are many others out there. It does emphasise the importance of approaching any new theory with caution. I believe that my four tests are helpful in this regard. As an aside, researchers at Tel-Aviv University used AI to examine 412 experiments designed to test four leading theories of consciousness.[75] These were GWT, higher-order thought theory, recurrent processing theory (not covered in this chapter) and IIT. They concluded that the experiments were largely, albeit inadvertently, designed to support the theory that the experiment was meant to be testing. As already proposed, any theory should be empirically falsifiable. In order to achieve that, experiments need to be designed to challenge theories in a rigorous manner.

Any new theory, therefore, including my own, can only ever be put forward with humility, recognising that it is speculative. In addition, any new theory of consciousness is highly likely to borrow elements and insights from other preceding accounts. Mine certainly does.

Chapter 12

Other Considerations with Respect to Consciousness

Recap

To recapitulate, to have consciousness is to make sense of the question, 'What is it like to be a …?' It is like something to be the President. It is like nothing at all to be a refrigerator. Consciousness is about having feelings, about experiencing. Feelings are what we attribute to ourselves when we answer the question, 'What is it like to be me?'

We have a difficulty, though. Normally, when we describe things, with the exception of abstract nouns, they have a physical presence. That seems to make the act of description feasible. Even abstract nouns are within reach. Words like honour and truth and justice have all been deployed with such frequency over the years that we more or less recognise them when we see them, and their imposters too. Consciousness does not appear to have a physical presence, although we intuitively feel it has a causative capability. Nor is there an accepted definition of it, even in one of the most obvious places to look, the world of neuroscience.[1] I can describe my consciousness but I cannot describe yours. We are each left with our own personal descriptions of consciousness and we can only hope that we are all talking about the same thing. So, we need to tread carefully.

Emotions are a good example of the need to be cautious when talking about consciousness. As explained in chapter 6, emotions are biochemical mechanisms, nothing more. To feel an emotion is to experience, or be conscious of, the biochemical mechanism at work. Yet, in common parlance, emotions are routinely conflated with consciousness: we talk about happy feelings. We should instead talk about feelings of happiness. Similarly, the production of a sentence takes place unconsciously, another mechanism.

One then becomes conscious of the sentence coming out of one's mouth. This is called speech and we are conscious of it. The same distinction applies to thought: we are conscious of thinking but we do not consciously think.

Even closer to home, vision, which we all take for granted, is another example of an area in which to tread carefully. When a camera sees a landscape, the photons reflected off each point of that landscape enter through the camera's lens. Processing inside the camera takes place in order to record digitally the landscape as seen by the lens. Similarly, when an animal without consciousness sees its immediate environment, the photons reflected off each part of that environment enter through the animal's eyes. Processing inside the animal's brain takes place in order to record neuronally the environment as seen through its eyes. When we humans, as conscious beings, see, the same biochemical process happens. In addition, though, we are conscious of what we are seeing. In other words, we additionally have subjective experience of the output from the biochemical mechanism of the conversion of the photons into neuronal processing.

Other aspects of consciousness include the feeling of agency: we feel as if we are the protagonist in the actions that we take. We decide what we are going to do, or so it feels. Consciousness appears, therefore, to be tied up with decision-making, as described in chapter 8. But the connection between the two is indirect. It is not as inextricable as it feels.

The recap above, though, leaves out apparently important features of consciousness such as wakefulness, awareness, attention and that favourite of philosophers, qualia, to which we will come. Moreover, the feeling of agency mentioned above does not appear to do full justice to the richness of the self. In that connection, we are also bound to look at the hoary old topic of free will. Finally, we are accustomed to thinking of consciousness in binary terms: either an organism possesses it or it does not. But, in possessing it, one might have more or less of it or be better or worse at deploying it. We shall look further at these topics below.

Wakefulness

As Antonio Damasio puts it, "Consciousness is not merely wakefulness."[2] In other words, wakefulness is a necessary but not sufficient condition of consciousness. Illustrating the point, Damasio provides a revealing description of a strange condition, "epileptic automatism".[3]

> In the middle of our conversation, the patient stopped talking and in fact suspended moving altogether. His face lost expression, and his open eyes looked past me, at the wall behind. He remained motionless for several seconds. ... When I spoke his name, there was no reply. When he began to

> move again, ever so little, he smacked his lips. His eyes shifted about and seemed to focus momentarily on a coffee cup on the table between us. It was empty, but still he picked it up and attempted to drink from it. I spoke to him again and again, but he did not reply. ... Finally he rose to his feet, turned around, and walked slowly to the door.

This account of epileptic automatism graphically illustrates the distinction between being awake and being conscious (i.e., having consciousness). In this case, the patient was awake but not, for a period, in possession of consciousness.

It could be argued that, when dreaming, one often has a sense of self and has awareness, albeit in a strictly limited sense. Thus, one has a form of consciousness when dreaming, that is to say, when one is not awake.

So, there are instances of being awake but not conscious and being conscious but not awake. Consciousness and wakefulness are different states, even if they come together most of the time.

Awareness

The term 'awareness' appears often in accounts of consciousness. It has in this book. For the sake of clarity, if nothing else, it is worth establishing whether awareness is a different thing from consciousness or merely a synonym. Incidentally, the sense of awareness we are considering here is subjective. We are not considering objective awareness in the sense of a motion sensor being aware of a burglar's entry into the room.

Chalmers summarises some of the issues inherent in this question in a passage entitled "The Principle of Structural Coherence".[4] This is the principle that the "*structure of consciousness*" is coherent with, i.e., effectively the same as, the "*structure of awareness*".[5] He notes that this does not mean that there are not some properties of consciousness and awareness that are different.[6] But there does not appear to be much in it. Many other authors take the same line, as shall I. In short, subject to the context, there is to all intents and purposes no difference between consciousness and awareness. A more fruitful area for enquiry is attention—how do consciousness and attention interact?

Attention

The same question posed in relation to awareness might be aimed at attention too. Is attention synonymous with consciousness? At first glance, it might appear that way. If I pay attention to something, then I tend to be conscious of it. If I am conscious of something, then I tend to be paying attention to, or attending, it. So, it would appear that consciousness and attention are the same thing because they always happen at the same time. Numerous researchers note, however, "that attention and awareness can be fully dissociated".[7] In other words, consciousness and attention are different things. They just happen to come together most of the time.

The answer to the conundrum lies with the unconscious mind. Thousands of sensory inputs stream into our brains all the time. Our heads would be spinning constantly if we focussed on each and every one of them, in other words if we registered each one consciously. We only need to, indeed only wish to, focus on the most important one or ones. The brain has evolved a system for achieving this.

Each of the inputs is unconsciously evaluated for relevance by a part of the brain called the basal ganglia.[8] In chapter 6, I introduced the ventral striatum, which is the most important part of the basal ganglia in the context of value-based decision-making. It is heavily involved in evaluation.

Using the basal ganglia, the brain compares different sensory inputs to determine which has the highest value. This determines which input is most relevant, i.e., to which one we should attend, on what we should focus. The process of evaluation performed by the ventral striatum weights each of the competing sensory inputs. A winner-takes-all mechanism then selects the winning input.[9] The evaluation and winner-takes-all processes are strictly unconscious. As Dehaene observes:[10] "It would be oddly inefficient for our mind to be constantly distracted by dozens or even hundreds of possible thoughts [or objects] and to examine each of them consciously before deciding which one is worthy of a further look."

Only when the winner-takes-all mechanism has produced the winning object does the brain then boost that attended object into consciousness. In other words, we become conscious of that to which we are attending. This is why attention and consciousness appear to go together. Dennett's multiple drafts model, as described in the previous chapter, is another way of describing the same process.

If this is true, then we should be able to identify examples of attention without consciousness. Damasio's example of epileptic automatism is one such example. His patient was awake and attended to the coffee cup but remained without consciousness. In fact, any state of wakefulness without consciousness is likely to signify attention without consciousness. Blindsight is another example. V.S. Ramachandran cites the discovery by the late Larry Weiskrantz, a professor of psychology, of this strange phenomenon in *The Tell-Tale Brain*.[11] Weiskrantz's patient, Gy, had sustained "substantial damage to his left visual cortex". This meant that he could not see out of his right eye. The left visual cortex supports the right eye visual field, and *vice versa*. Despite the fact that theoretically Gy was not able to point to things available only to his (non-working) right eye, i.e., not visible to his (working) left eye, he could. As Ramachandran describes it:

> In the course of testing Gy's intact vision, Weiskrantz told him to reach out and try to touch a tiny spot of light that he told Gy was to his right. Gy protested that he couldn't see it and there would be no point, but Weiskrantz asked him to try anyway. To his amazement, Gy correctly touched the spot. ... repeated trials proved that it had not been a lucky stab in the dark; Gy's finger homed in on target after target.

Subsequent research has persistently supported this finding. The reason remains the subject of debate and research. Suffice it to say, there are various visual pathways in the brain. In certain cases, damage to one of these pathways creates blindsight in the patient. In other words, the patient is no longer able to see consciously in the relevant field of vision but successfully attends to items in that same field of vision anyway.

What about the sharply focussed attention of a lioness on a succulent looking baby warthog—does she possess consciousness? We shall look at the highly-charged question of non-human animal consciousness later in the chapter.

Psychologist Michael Cohen *et al.* conclude that attention without consciousness is indeed well supported in research.[7] Attention and consciousness are different things. Indeed, it remains unclear why most organisms would need consciousness given the foregoing description. Yet they certainly need to be able to pay attention to their environment. That is precisely why organisms possess sensory organs. Conversely, Cohen *et al.* conclude that there is no empirical evidence yet meeting the criteria of awareness, i.e., consciousness, without attention.

Working memory

Working memory, like attention, is sometimes confused with consciousness. An example of this would be the short-term recall of a telephone number whilst entering each digit in turn on to your telephone keypad. It is easy to see why the confusion might arise. One's recall of the first digit in the telephone number is experienced consciously. The same thing happens with the second digit, and so on. Working memory and consciousness go together. Perhaps, therefore, they are the same thing, or so the argument would go. But, as the neuroscientists Bernard Baars and Nicole Gage note,[12] working memory has "unconscious aspects" too, even though these may be seconds, or even fractions of seconds, away from becoming conscious.

Imagine performing the calculation 259 times 63 in your head. Your working memory retains those digits throughout the calculation. Specific digits, however, drop out of conscious working memory while you are working on other digits. For example, you might decide to multiply 259 by 60 prior to multiplying 259 by 3. You will then add the two products of your multiplication together to derive the final answer. While you are working on 259 times 60, the 3 will drop out of your conscious working memory. It will be held waiting in unconscious working memory for immediate retrieval once you move to the second stage of the calculation.

Like attention, working memory and consciousness can be, and should be, dissociated.

Self

Chapter 11 noted the distinction drawn by Antonio Damasio as between Self-as-object and the outside world. Self-as-object essentially comprises all the neuronal circuitry representing the body. In his account, the brain represents both the outside world and also the objects we perceive in it in a separate array of neuronal circuitry. Meanwhile, Self-as-subject emerges out of Self-as-object in the form of a further neuronal network. Consciousness occurs as the Self-as-subject network interacts with the brain's representations of the objects we perceive all around us.

An alternative reading of Self-as-subject and Self-as-object might ascribe almost all the neuronal circuitry in our brains to Self-as-object. Self-as-object's content, so to speak, would then comprise the neuronal representations of both our bodies and also the objects all around us. Self-as-subject would constitute a separate neuronal network. It would not emerge from Self-as-object. Rather, its content would come to it from Self-as-object. The question in this alternative reading is, from where does Self-as-subject arise? Instead of looking inside the brain for the answer to that, perhaps the answer lies outside. Perhaps Self-as-subject is imported from others.

Daniel Wegner, a professor of psychology, explores exactly this idea through his concept of the "virtual agent".[13] The brain naturally creates mental representations of both other people and of itself. To all of these mental representations is attributed agency; hence, the term, "virtual agent". We see someone do something and we conclude that the someone is an agent, an entity that causes things to happen. If y follows x, our brain does not just assume that at one moment x happened and then at another moment y happened. It assumes that x caused y. If an antelope failed to grasp that a lion caused the death of its companion, then it would never learn that running away from a charging lion in the future was the sensible course of action.

Wegner proposes the idea of a "virtual agent" piggy-backing on, in this case, a collection of memories. To memories we could add perceptions as well. He seems to suggest that the brain perceives or infers a virtual self in others. In similar fashion, the brain adds or instantiates a virtual self within its own architecture. This would be Self-as-subject. Self-as-object would supply the content. As Wegner observes,[14] "The development of an agent self in human beings is a process that overlays the experience of being human on an undercarriage of brain and nerve connections." I will come back to this idea in chapter 15. In the meantime, Wegner's reference to the term 'agency' leads neatly into the next section, whether consciousness is causal.

Causality

I first touched on this question in the previous chapter. Chalmers produced an analysis of different approaches to consciousness, distilled into six types, A–F. Only type-E consciousness holds that consciousness is not causal. Type-E consciousness is called epiphenomenalism. In brief, it holds that subjective experience has no causal impact on our brain or our behaviour. It is a challenging position to take. It certainly breaches the evolutionary use test (test four). If consciousness cannot cause anything, including the brain, to happen or make things happen, then what is the point of it? The neuroscientist Michael Graziano points out another flaw in the epiphenomenalist position:[15] "Awareness must be able to act on the brain, to supply the brain with specific, reportable information about itself, or else we would be unable to say that we have it." In other words, if consciousness is not causal then it could not cause us to say anything about it, and yet we do.

We have a conundrum. On the one hand, consciousness looks as if it is causal in some way. On the other hand, Self-as-subject, where consciousness seems to reside, seems to be shadowy and insubstantial. In fact, it looks as if Self-as-object generates all the content of which we are aware. So, it seems as if there is no way by which consciousness can exercise its ability to be causal.

Wegner attempts to resolve this conundrum by proposing that consciousness is a useful illusion. The "illusion" aspect of this description is rendered into a theory.[16]

> The theory of apparent mental causation, then, is this: *People experience conscious will when they interpret their own thought as the cause of their action* (Wegner and Wheatley 1999).[1][17] This means that people experience conscious will quite independently of any actual causal connection between their thoughts and their actions.

The *useful* aspect of this description is the proposal that the feeling of conscious will underpins human morality by giving weight to relevant emotions.[18,19,20] This enables the theory to pass the evolutionary use test.

In other words, consciousness is not directly causal: it does not directly cause actions. So, if it is not directly causal, then it must instead possess an indirect causality. An example of indirect causality might go as follows. Imagine a small boy doing something naughty in his room such as using his colouring set of indelible ink pens to draw all over the walls. He looks up to find his mother standing in the doorway watching him, expressionless. The boy stops immediately. It is certainly true that, without his mother's sudden presence, the boy would not have stopped. Yet his mother said nothing, did nothing and indeed did not even make a face. She did nothing that could be described as directly causative of the boy's stopping. In fact, it was the boy's own imagination of admonition or punishment that directly caused him to stop. His mother's appearance was only indirectly causative.

I suggest that it is precisely in this grey area of indirect causality that consciousness operates. This is one of the reasons that consciousness is so difficult to pin down. It seems to be neither one thing, nor the other. My theory of consciousness in chapter 17 builds on this idea.

Free will

The discussion of causality provokes the question, where does that leave free will? Free will is a subject beloved of philosophers, not least because one can go round and round in circles trying to pin it down. It comes down to the feeling of agency that we have when we decide to behave in a certain way or take a certain action. In other words, we feel that we have caused the behaviour or the action. That is, supposedly, our free will. It is intimately connected with consciousness. If you did not have consciousness, then you would not have the feeling of free will. With consciousness, though, you feel that you have free will.

That does not mean that they are the same thing. Consciousness is a state. Free will is a faculty that we feel we possess. Specifically, it is the feeling that I myself, my conscious self, can cause things to happen. But we have already established that consciousness is only indirectly causal. We must infer that free will in the strong sense of the phrase does not exist. Rather, free will just feels as if it exists in that form. It might instead exist in a weak sense, reflecting the indirect nature of consciousness's causality. Consciousness might, for example, indirectly cause us to reflect before automatically taking a certain action. This is not to revert to Libet's conscious veto, underpinned by a conscious mental field. This would be another type of mechanism. We shall need an explanation of that sort of free will.

There are two other avenues down which discussions about free will can go. One is to suggest that free will is solely to do with freedom from constraint by an external agency. Whilst free will no doubt has to be free in that sense, that is fairly clearly not what is at stake here. The other avenue invokes the possession of a soul or spirit of some kind. I have already noted that I am setting such ideas aside for the purposes of this book.

Consciousness in non-human animals

The obvious difficulty with discussing non-human animal consciousness is that non-human animals cannot talk to us. The animals of each species tend to communicate with each other at some level but they do not use any of the human languages to do so. So, they cannot say the sorts of things that we humans say when speaking to each other that in turn leads us for our part to infer that all we humans possess consciousness. This creates a real problem in analysing consciousness more broadly. It would be easier to nail down consciousness if we knew precisely which animals had it and which did not. Then, we could in principle, by comparing brains, figure out where and why it evolved.

That does not stop many people anthropomorphising their pets and thereby concluding that many non-human animals do indeed possess consciousness. This is despite much evidence to the contrary. For example, many dog owners believe their dog is capable of feeling guilty on the strength of their apparently guilty look when they have done something wrong. Note, I have used the expression 'feeling guilty' quite deliberately. It means being conscious of the (underlying) emotion of guilt. In studies done in 2015, researchers in the USA and Hungary concluded that[21]

> dogs have developed a submissive behaviour tactic that is not based on the knowledge they have done wrong but is an 'appeasement response' to their owner's anger. They believe it is an evolutionary strategy that has helped dogs to live more easily in close contact with human beings.

In other words, there is no feeling of guilt at all. In fact, it would appear that there is not even the underlying emotion of guilt. Submissive behaviour is common within animal groups in the wild. The difference is that humans perceive such doggy behaviour in the home. They then impute the feeling of guilt to the animal. This is not to say that such behaviour absolutely and definitely does not follow the feeling of guilt. But the simpler and more compelling explanation is that it has nothing to do with guilt. Rather, it is just a learned, or even evolved, behavioural trait. This is also not to say that dogs do not have emotions. They do: we can trace the pathways their emotional circuits take through the brain. As described earlier, however, there is a distinction between emotions and feelings of emotions. The bar for imputing the latter in a dog, for example, must be much higher. After all, we still have no idea where such feelings might be located in the brain. So, in this area, as in so many areas to do with consciousness, we must tread carefully.

Some scientists believe that consciousness originated in non-human animals a long time ago. For example, Professor Derek Denton, a leading physiologist, believes that[22]

> the imperious states of arousal and compelling intentions to act that characterize *the primal emotions* were the origins of consciousness. The term 'primal' or primordial emotion is being used for the *subjective element* of instinctive behaviour...

In referring to the primal emotions, Denton is referring to such things as "thirst" and "hunger for air". Peter Godfrey-Smith builds on this theme, calling this type of early consciousness "subjective experience" or "simple experience".[23] He dates this to the Cambrian era (around 500 million years ago), when animals with nervous systems exploded in both number and variety. Other scientists, Gerald Edelman and Jaak Panksepp, for example, do not carve out subjective experience in quite this way. They propose that "primary consciousness" (Edelman[24,25]) or "affective consciousness" (Panksepp and Biven[26,27]) originated with the appearance of mammals and birds (broadly, some 200 million years ago).

Denton, Godfrey-Smith and Panksepp are all effectively saying that the organism's subjective experience or feeling is what lends the relevant emotion its driving force. This is why Denton uses the adjectives "imperious" and "compelling". Test four (evolutionary use) is clearly not failed in such theories. But their proposals appear to breach a variant of test two (lo and behold): emotional drives compel behaviour and, lo and behold, subjective experience has clearly emerged. Edelman recognises this danger.[28] So, although he concludes that the relevant neuronal processes lead to consciousness, he also proposes that consciousness itself is not causal. Thus, only the underlying neural basis of consciousness is causal. It could be argued that this approach softens the lo and behold problem, albeit only in the sense that it no longer really matters. Of course, this approach potentially takes him into the strange space that is epiphenomenonalism.[29]

Suffice it to say, non-human animal consciousness remains both an uncertain and a highly-charged area of debate.

Qualia

In philosophy, the subjective experience of the redness of an apple is termed a *quale*; the plural is *qualia*. The challenge posed by many philosophers is: how do we account for all these different qualia? In other words, how do we account for the subjective experience of the redness of that apple, the greenness of this apple, the blueness of that bit of sky and the greyness of this other bit of sky? In fact, this is not just about colour. An apple also appears smooth and round, each of which constitutes further qualia. But the curious thing is that, if one looked at an apple with a scanning transmission electron microscope, one would not see any of those features. For example, one would not see redness at all. The apple is not red at the point of departure, so to speak. Yet, it is red at the point of arrival; that is, in your vision. As Dennett notes:[30] "colours were made to be seen by those who were made to see them."

In other words, organisms evolved the equipment needed to turn certain types of sensory input into red at the point of arrival, for example. So, we can re-evaluate the question, did consciousness generate all those qualia? That would place a heavy burden on consciousness. The redness of an apple is the result of the interaction of two systems, one on transmit and one on receive. The quale of the apple's redness is the subjective experience of the redness. Consciousness did not make the apple red. It was red once the photons from the apple had hit your eyes and then sparked a neuronal chain reaction leading back to the visual cortex and forwards again.

This means that objects have certain properties that only exist when there is an observer. The philosopher John Locke called such properties "secondary qualities", to distinguish them from "primary qualities" such as height and width and shape, for example.[31] An apple is not properly red until someone observes the apple and sees that it is red. The redness of the apple is one of its interactive, or shared, properties. If evolution had waited around until the emergence of consciousness for the emergence of shared properties such as redness, then there would have been plenty of non-conscious organisms left bereft. For, without such shared properties, they would have been unable to make much sense of their environment.

All organisms react to the shared properties of things in this way. Let us take insects, most of which have colour vision. Let us assume that insects do not possess consciousness. What then does the redness of an attractive flower mean to an insect? We are driven to two conclusions. First, the insect can 'see' the flower's redness. Secondly, the insect is not aware, or conscious, of that redness. Whilst surprising, counter-intuitive even, it should be no more surprising than the phenomenon of blindsight described earlier.

This, along with some of the other points made above, seems to suggest that consciousness is parasitic. Indeed, our continuing inability to locate the neural correlates of consciousness in the brain implies such a relatively minor but diffuse physical presence for consciousness that neuronal parasite might be the best description for it. Consciousness does not put the redness into red. It does not put the niffiness into smells. It has evolved in the brain to sit on top of redness and smells, not to give them their efficacy.

Degrees of consciousness

Damasio introduces two types of gradation of consciousness. The first is an "intensity scale", which he illustrates by referring to the in-between state we experience as we wake up, for example.[32] The second is the distinction he makes between "core consciousness" and "autobiographical consciousness",[33] as described in chapter 11. I am instead interested here in a third type of gradation of consciousness. Specifically, this is to do with whether the consciousness that we humans possess (Damasio's autobiographical consciousness) can present with different levels of skill or utility.

Bearing in mind Searle's point that consciousness will have an evolutionary use or purpose, different people are likely to be better or worse at it. There is no particular reason why consciousness might not be like any other capability we possess. Take fleetness of foot: some people are quick runners and others are slow. Take intelligence: some people are bright and others are not. The probability is that consciousness also comes in different degrees or standards in terms of people's facility with it.

Chapter 13

Summary of Key Points from Part 3

Part 3 has introduced the topic of consciousness in some detail, setting out in particular what consciousness is not.

1. Thought is not consciousness, which is in turn not language. The three of them often go together but they are distinct. In particular, thought is unconscious. It has its own language, mentalese.
2. Mentalese is translated into ordinary language by the language areas of the brain. We hear the sounds of that language and we feel them as meaningful. But the meaning itself is unconscious. Rather, words appear in our heads or come out of our mouths. So, we become conscious of language rather than consciousness producing language.
3. Thought evolved before consciousness. Many, maybe all, non-human animals possess thought. In turn, consciousness probably evolved before language, or at least before fully-fledged language.
4. Consciousness needs an interface in order to optimise its work with thought. Language is that interface. A fully-fledged language makes a huge difference to the ability of consciousness to perform its role. Our language abilities have given a strength to our consciousness unmatched in the rest of the animal kingdom. Consciousness, through the medium of language, hones and enriches our thought by promoting our Type 2 processing. It turns out that thought, language and consciousness stand in a virtuous circle. In short, we can control ourselves better than any other animal and thereby unleash a greater emotional and intellectual power than all other living things.
5. In "What is it like to be a bat?", Thomas Nagel asked the question, 'is it like something to be a bat?' Whilst it is not at all clear that

bats possess consciousness, this formulation of what it is to have consciousness is intuitively sensible. So, setting aside bats, one might ask, 'What is it like to be the Prime Minister or the President?' Such a question gives rise to a raft of possible answers, all of which will probably make sense. But, ask the question instead, 'What is it like to be a refrigerator?', and there is no answer that makes sense. It is not like anything to be a refrigerator. A refrigerator just is. In other words, a refrigerator does not have consciousness.

6. David Chalmers asked why consciousness needed to be, or feel, like anything at all. This is the so-called hard problem of consciousness. It is an important challenge and it leads to the question, 'What does consciousness bring to the party?'
7. I introduced the notion of four tests that might be applied to any theory of consciousness. The homunculus test exposes any attempt by a theory to slip in a homunculus, a little person in the brain, to give consciousness its force and efficacy. The lo and behold test asks a putative theory to explain how a set of physical processes might give rise to the apparently intangible feature of consciousness. The conscious thought test says that all thought takes place in the unconscious mind and insists that any putative theory meet that challenge. The evolutionary use test asks a theory to explain why consciousness might have evolved and then persisted.
8. For these purposes, consciousness is not the same as wakefulness, although subject to certain dream states one needs to be awake to be in a state of consciousness. Similarly, for humans, attention and consciousness seem generally to happen at the same time but it is perfectly possible to conceive of attention without consciousness. Indeed, this might be the natural state for most animals.
9. Consciousness is not directly causal: it does not directly cause actions. But test four tells us that consciousness must serve a purpose to the extent that consciousness is not held to be, for example, ubiquitous in nature such as is deemed to be the case by panpsychism. So, if it is not directly causal, then it must instead possess an indirect causality. I suggest that it is precisely in this grey area that consciousness operates.

Part 4 will propose a means by which consciousness came to be instantiated in mankind's antecedents and, thus, in *Homo sapiens*. It sets out my own theory of consciousness. Crucially, the theory is rooted in the use or purpose of consciousness. It is of necessity speculative but my aim has been to let the facts, as set out in parts 2 and 3, dictate the parameters of the theory.

Part 4, therefore, lays the groundwork for instantiating consciousness in a robot. As noted earlier, the aim is not to emulate the purpose of consciousness for humans in AI. That purpose is primarily about decision-making optimisation. Instead, the aim is to emulate subjective experience in AI so that AI, or rather ASI, fully empathises with us. Only in that way can we be optimistic that ASI's proposed solutions to our intractable problems are best suited to our interests.

Part 4.

The Mirrored Homunculus Theory

Chapter 14

Decision-Making and Complexity

When we left off the discussion of decision-making at the end of chapter 8, we had reached the point of positing a biasing input in the brain. I suggested that the stochastic nature of our decision-making required a biasing input to help achieve emotional reliability. (By way of reminder, a stochastic system operates randomly. It is, however, possible to estimate the probabilities with which different outcomes will emerge from the system.) We need to be able to act on the basis of our social emotions in preference, in many cases, to our primary emotions. We also need such a biasing input to give Type 2 processing time to work. A biasing input is needed to put Type 1 processing back in its box in order to allow Type 2 processing to keep working at a novel or complex problem.

I propose that consciousness is that biasing input. We will come to ways in which that hypothesis might be tested in chapter 18. In the meantime, I propose that consciousness evolved to enable its possessors to help address the greater complexity of the advanced social groups in which great apes and certain cetaceans came to live. It also evolved to sustain Type 2 processing to help tackle difficult problems and decisions.

I believe that the evolution of consciousness represented such a fundamental change in decision-making that one can picture it as a discontinuity in the animal kingdom's decision-making continuum. It is for this reason that consciousness persisted once it had evolved. It added reliability and sustainability to our decision-making processes. This allowed us to behave prosocially in complex societies and to tackle novel and complex problems.

This may still not be intuitively obvious. Why should value-based decision-making (VBDM) not cover the full spectrum of decision-making from a rat freezing at the sight of a predator to, say, troops in the First World War charging headlong over open ground into sustained machine-gun fire? That is, after all, the beauty of VBDM theory—its ability to evaluate different possible actions and then select a winner from them. On the face of it, it accounts for a large part of the continuum of decision-making complexity. I referred in chapter 8, however, to the fact that it remains an open question how the brain tackles more complex questions. This chapter delves into the detail of different types of decision along the continuum. It points to the outline of a discontinuity in the continuum. At that discontinuity, a new ingredient needed to be added to the mix in order for VBDM to retain its efficacy and plausibility. I propose that consciousness evolved to be this new ingredient.

Frightened rats

As Joseph LeDoux notes, "Laboratory-bred rats who have never seen a cat will freeze if they encounter one."[1] Freezing at the sight of a predator comes naturally. For a start, predators are particularly well-equipped to perceive movement. So, the rat's freezing makes sense. Moreover, it gives the potential victim time to prepare for the next step, i.e., flight or fight.[2] This type of behaviour is instinctive, or reflexive.

But similar responses to changes in the environment can be learned too. Such learning is called *Pavlovian* after the Russian physiologist, Ivan Pavlov. Pavlov became famous for his pioneering work in the field of behavioural conditioning. Certain stimuli, e.g., the cat, trigger certain behaviours, or responses. But it is also possible to train or condition animals, including humans, to respond to new stimuli. For example, a bell might be rung to signal the imminent arrival of the cat. With repeated bell–cat combinations, the rat will learn, i.e., be conditioned, to freeze at the sound of the bell. The rat will not wait until the cat appears. As Nathaniel Daw and John O'Doherty observe,[3] such "behaviours are more flexible than simple reflexes" because the stimuli can be made to vary through a learning, or conditioning, process. But, critically, "the responses themselves are stereotyped": the rat freezes in each case.

The rat is capable of evaluation (phase 1 of VBDM) but the amount of action selection (phase 2 of VBDM) is minimal. The rat just freezes every time. Phase 3 (learn) is the interesting phase in this case. The rat has learnt, through prior conditioning, about the significance of the bell. The emotion of fear is the main trigger for the rat's behaviours, the amygdala being central to the fear emotion network. We can, as a result, trace pathways in the brain from emotions to decision-making, even if the rat's level of decision-making is not complex.

Incidentally, it is important to emphasise the automatic nature of these processes. Even plants appear to work in similar ways. Evolutionary ecologist Monica Gagliano discovered this by working with pea plants. She used an established conditioning methodology.[4] She persuaded the plants to respond to an electric fan that had previously been associated by the plants with light (see figure 14.1). Light to plants is as attractive as catnip to cats and plants grow towards it, as illustrated in the left-hand column of figure 14.1. In the middle column, plants learn that the fan is a precursor to the subsequent appearance of light. This is the conditioning phase. The right-hand column illustrates the fact that the fan is now enough on its own to cause the pea plants to open up. Gagliano also showed that the plants could remember the significance of the fan weeks later. One could conclude, fancifully, that plants are, therefore, sentient. A more parsimonious explanation is simply that learning and memory are not such special tricks after all.

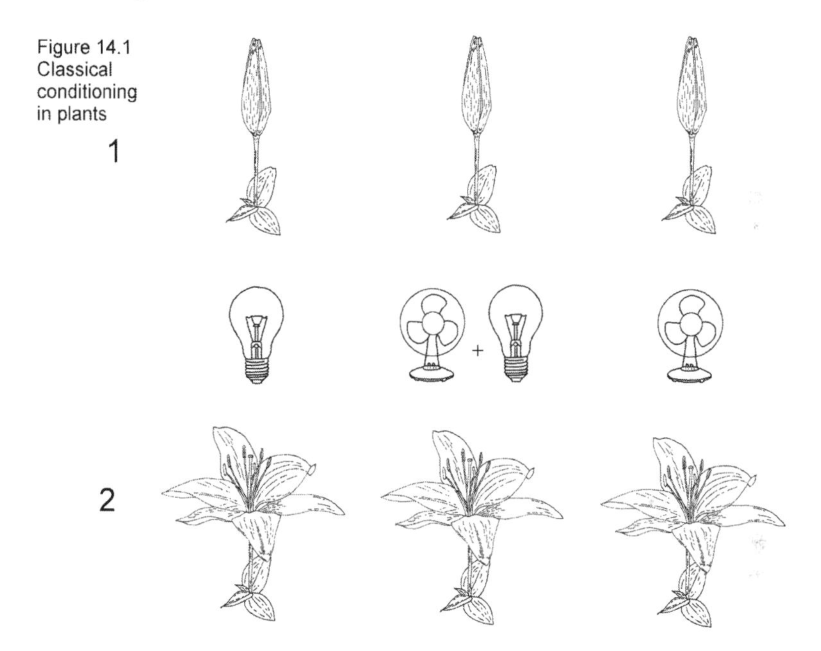

Figure 14.1 Classical conditioning in plants

The next level up in decision-making processes is termed *"habitual learning"*.[5,6] At this level, the animal can be conditioned to learn non-reflexive responses. When, for example, a dog has been trained, or conditioned, to walk on its hind legs if its owner's finger is raised (e.g., to signal a treat), that is a non-reflexive response. A dog in the wild does not walk on its hind legs. This is interesting for our purposes here in that the action selection processes of the brain have to do more work. The responses are not stereotyped. As a result, the complexity of the brain's processing at this level has increased. We begin to see real differences in behaviour, but they are engendered in an automatic way. As Daw and O'Doherty explain:[7] "such a mechanism is incapable of evaluating novel actions (or re-evaluating previously experienced ones) based on any other information about the task, the world, or the animal's goals."

So far, then, we have not encountered any behaviours that appear to rely on consciousness. Novel behaviours, though, may. What is needed for behaviours involving novel actions is "a *goal-directed* mechanism that evaluates more prospectively, as by a cognitive map".[8] The animal in question needs to be able to figure things out on its own. Let us turn to one of the cleverest animals of all, the New Caledonian crow, to elucidate this more clearly.

We should note in passing an important qualification to any conclusions or hypotheses drawn from the study of animal behaviour. Since animals cannot talk to us, inferring what is going on in their brains or minds is not easy. A recent experiment with another corvid, the Eurasian jay, illustrates the point.[9] Elias Garcia-Pelegrin subjected a jay called Stuka to a number of tricks to see how it coped. One such trick, the fast pass, is a trick that frequently fools humans, but birds possess a natural ability to follow rapid movements. Elias started with a worm in one hand. He moved the worm quickly to the other hand and then invited Stuka to indicate in which hand the worm lay. Stuka guessed wrongly. It did not see the worm being passed to the other hand. Garcia-Pelegrin inferred that, because most birds prefer to use one eye at a time, Stuka had been unable to switch quickly enough to binocular vision instead. All this can be, however, is an inference. However smart Stuka is, it cannot talk. Moreover, not all the animals in our experiments are necessarily configured in the same way. For example, we cannot be sure that birds' brains work in the same way as mammals' brains. So, with those caveats, let us do our best to consider the cognition displayed by the New Caledonian crows in more detail.

Crows and cognition

All corvids are intelligent by non-human animal standards but New Caledonian crows stand out as being amongst the brightest in the corvid family. An experiment using these crows established that their intelligence even extends to making, not just using, compound tools.[10] These are tools composed of more than one element.

The researchers deployed eight crows in their experiment. One by one, each crow was confronted with a partly transparent test box and various implements (see figure 14.2). The task was to push or drag a little trolley containing food sideways along a rail within the test box. At the end of the rail was a hole in the side of the test box, out of which the trolley would drop. The crow could then walk round the side of the box and eat the food from the trolley. The researchers provided the crows with a small variety of sticks and stick-like implements to enable the crows to push the trolley. To do this, the crows had to insert a stick through a wide slot along the bottom of the box and push the trolley from left to right along the rail. The challenge for the crows was that none of the sticks and stick-like implements was long enough to reach the trolley on the rail from the slot in the test box. They needed to be combined to make a long enough stick to reach the trolley.

Figure 14.2

Illustration of experimental set-up

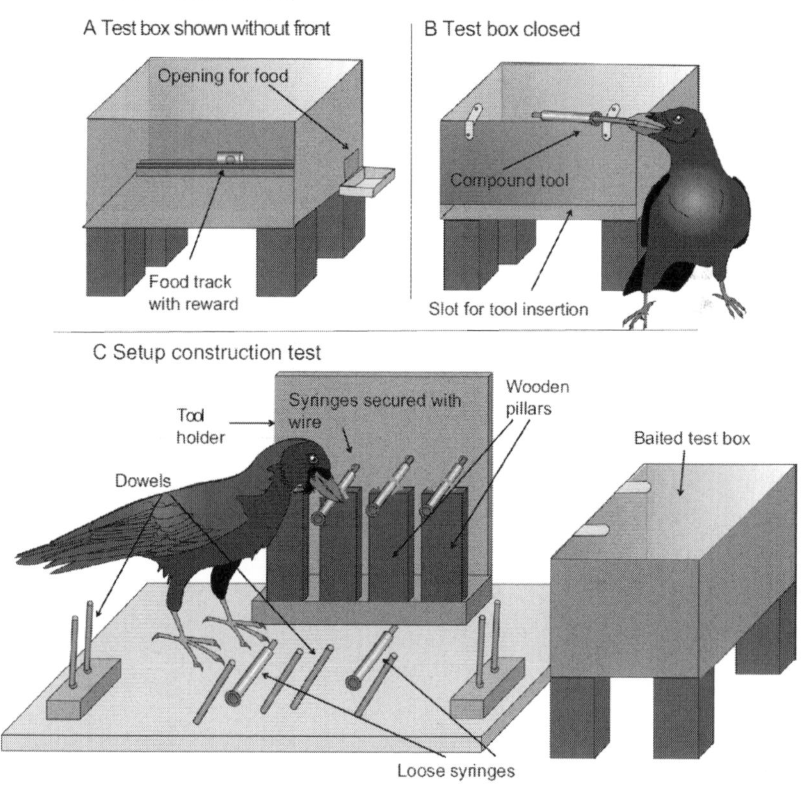

Figure reproduced in black and white from figure 1 of research paper,[10] under https://creativecommons.org/licenses/by/4.0/.

After a session that allowed the crows to familiarise themselves with the test box, the rail, the trolley, etc. by using an extra-long stick or dowel, the crows were tested one by one. There were four tests. Four out of eight crows passed the first test. This was to construct a long enough stick out of two shorter sticks that were designed to be able to be joined together. The combination stick was then long enough to push the trolley and secure the food on it. The subsequent tests became more difficult. One crow, Mango, managed to pass all four tests. The fourth test obliged

Mango to construct a stick of the requisite length out of three and then four shorter sticks.

Translating this into VBDM terms, each crow needed to experiment with the tools at its disposal in order to learn or work out the answer. It needed to cycle through the three phases of VBDM a number of times. This process entailed both emotion and cognition. The principal emotional drive would have been hunger. It might be argued that Panksepp's SEEKING emotion (see chapter 6) was also at work, impelling the crows to experiment. The cognition involved is more "opaque", to use the researchers' own term.[11] It is clearly Type 2 processing. But it is not clear how it worked—in other words, how the crows learned how to join small sticks together to make a compound stick that was long enough.

We shall start with phase 1 of the VBDM process. From similar experiments involving rats, albeit significantly less challenging in cognitive terms, Daw and O'Doherty conclude that cognition in such cases is underpinned by a goal-directed mechanism in the brain. They equate such a mechanism with "model-based reinforcement learning".[12] I described this in chapter 8. Model-based learning relies on mental representations, or models, of possible outcomes. Put another way, the brain needs to have a mental picture of what things will be like if it decides to take one action. Moreover, it needs another mental picture of what things will be like if it decides to take some other action. It needs to be able to simulate these possible outcomes. Each of these model futures then needs to be evaluated, so that comparison is able to take place and a decision or choice can be made between them. The difficulty is that, whilst model-based learning seems to hold water conceptually, we do not know how the brain actually realises it at the neuronal level.

In a similar vein, the researchers using New Caledonian crows observe that we cannot yet devise a model of the crows' learning process in operation.[13] This goes back to a problem that we came across in chapters 1 and 8. AI research has still not cracked the problem of answering questions without the support of large quantities of data. Machine learning relies on such data to generate its algorithms. Novel situations leave it cold. Unfortunately, we still do not understand in detail how the brain addresses such situations either. Model-based reinforcement learning is part of the answer but we do not know how the brain constructs, maintains and utilises the model in question.

By comparison, explaining how the crows achieved phase 2 of VBDM in the process, i.e., action selection, is easier. A winner-takes-all, stochastic action selection process would more than adequately explain actions selected by the crows.

The other question is what impelled the crows to persist with Type 2 processing rather than reverting to Type 1 processing. Type 1 processing is much less effortful but in this case it would have yielded no food. Type 1 processing is highly unlikely to have led Mango to make an extra-long stick out of three or four shorter sticks. The answer lies in part, at least, with the emotions involved. The main emotional driver was hunger: the crows wanted the food. Panksepp's SEEKING emotion might also have been present. Was consciousness also at work, providing a biasing input of sufficient heft to tip the crows into Type 2 processing? It is difficult to tell. This is a good example of why it has so far been impossible to prove whether or not non-human animals possess consciousness.

Let us now turn to the complex decision-making challenges entailed by living in complex societies.

The social life of a baboon

The late Dorothy Cheney, a professor of biology, and her husband Robert Seyfarth, a professor in animal behaviour, have written the definitive book on baboon society: *Baboon Metaphysics*.[14] Much of their book focuses on a group of chacma baboons, one of five species of baboon. Over the period of study, the group averaged 80 individual baboons, with adult females accounting for just over a quarter of this number. The number of adult males is less than half the number of adult females. Typically, a male baboon will grow to adulthood and then subsequently leave to join another group elsewhere.

The adult females represent the core of the group. Female ranking is complicated. The group is composed of a number of families, each with a female at the top. The families themselves are rank-ordered. Within each family, there is in turn a ranking of each family member. Female baboon interactions are sophisticated and nuanced both within and between ranks and families. They involve favours given and received, friendships, alliances and rivalries. Physical interaction includes a wide range of vocalised communications, grooming and the odd slap or scuffle. Female interactions generally avoid the blunt instrument of combat exhibited by the males. They are also made more complex by the presence of many juveniles in the group. The question is, how does each female keep track of all this ranking and the other attendant behaviours? The mental calculations involved seem daunting. Cheney and Seyfarth conclude that the baboons each perceive connections between their behaviour and the responses by other baboons to that behaviour.[15] This is a "simple learning mechanism" and it "provide[s] powerful and often accurate means for animals to assess the relationships that exist among others". But is it really enough to account for the huge complexity of these baboon groupings?

We are still not clear as to how in detail the baboon's brain tackles the problem. Of course, part of the answer may be that it only partially tackles the problem. Maybe the baboon's brain is more or less capable of remembering the various relationships within the group but is only partly able to compute with that knowledge. A baboon may know that, if it grooms another baboon, then a certain third baboon may become annoyed. But it may not be able to work out what the impact will be on the fourth, fifth and sixth baboons also present.

Another dimension to the understanding of how baboon society functions concerns theory of mind. Theory of mind is the attribution of mental states to others twinned with an idea as to what those mental states are at a given moment in time. So, for example, my own theory of mind allows me to imagine what you might be thinking or intending to do. It can only ever be an attribution because we cannot directly observe the mental states of others. We can only observe our own mental states and, even then, only to a point. If baboons possessed theory of mind, would that indicate that they possessed consciousness?

We are so used to thinking of intentions and choices as being part and parcel of consciousness, because we ourselves are conscious, that we forget that this might not necessarily be the case. Cheney and Seyfarth conclude that:[16] "Baboons' theory of mind might best be described as a vague intuition about other animals' intentions." But they do not know if this is mere perception of intent in the same way as Pepper can recognise a small number of human emotions. The recognition of others' intentions could simply be unconscious contingency-based learning. Put another way, the recognition of intentions in others could just be an automatic process based on experience.

The alternative would be perception of intent combined with a conscious awareness of it. There are accounts of baboon interaction with humans that suggest that the way that baboons think about others goes beyond simple learning mechanisms and vague intuitions.[17] They seem to be acutely sensitive to human posture and behaviour, which might or might not lift baboons above such lesser developed facilities. Once again, it is difficult to tell whether consciousness is involved, or needs to be involved.

Chimpanzees appear to be more behaviourally sophisticated than baboons. Their groups are less stable and, therefore, more complex. Relationships are constantly shifting, as indeed is the membership of the group. As Cheney and Seyfarth put it, they live in "fission–fusion communities".[18] This "may place strong selective pressure on the ability to imagine and plan hypothetical social interactions with individuals who may not be encountered for several days or even weeks".[19] This echoes

Frans de Waal's story, related in chapter 7, of a chimpanzee avenging itself on four of its peers days after having been wronged. Cheney and Seyfarth conclude that[20] "chimpanzees' social structure may have favoured a language of thought that includes a form of 'what if' episodic memory that is qualitatively different from that found in monkeys." (A baboon is a monkey.)

Some chimpanzees also use relatively complex tools in the wild such as the combination of a large stone and a small stone to crack nuts. The chimpanzee places the nut on the large stone and cracks it open with the small stone. Recent research, however, concludes that chimpanzees do not in general innovate nut cracking.[21] Rather, a chimpanzee community was probably lucky enough to start nut cracking once by virtue of a chance discovery or the efforts of an exceptional chimpanzee.[21,22] Subsequently, so-called high-fidelity social learning then enabled the discovery to be passed down through the generations in that community. Indeed, another community of chimpanzees only six kilometres away does not engage in nut cracking.

None of the above proves whether the baboon's or chimpanzees' decision-making has the benefit of conscious input. We should not be downhearted about the fact that there is no obvious bright-line rule[23] to distinguish between where VBDM theory can plausibly explain the brain's decision-making processes and where, on its own, it cannot. I noted earlier in the chapter that there appeared to be a discontinuity in the decision-making continuum. At this discontinuity, a new ingredient needed to be added in order for VBDM to retain its plausibility. I proposed that this new ingredient was consciousness.

For the time being, we will have to be content with speculating that the behaviour of some species might be relatively complex but is still just explicable by stand-alone VBDM theory. The baboon might be such a species. By the same token, there might be (a smaller number of) other species aside from humans whose behaviour is so complex that they find themselves on the other side of the discontinuity. Maybe the chimpanzee is such a species. Maybe crows fall into that category too. The trouble is that without language we cannot discuss the matter with them. Nor are we able to locate the neural correlates of consciousness. So, we cannot yet draw the bright-line between the two types of non-human animals. Fortunately, there are many instances of human behaviour that demonstrate how the proposed discontinuity of consciousness operates.

The Charge of the Light Brigade

> "Forward, the Light Brigade!"
> Was there a man dismayed?
> Not though the soldier knew
> Someone had blundered.
> Theirs not to make reply,
> Theirs not to reason why,
> Theirs but to do and die:
> Into the valley of Death
> Rode the six hundred.

So go the words of verse two of Alfred, Lord Tennyson's poem, *The Charge Of The Light Brigade*. The poem was written in late 1854 about a particularly bloody and foolhardy episode in the Crimean War. Orlando Figes described the sequence of events leading to the fateful charge in his book, *The Crimean War*.[24] On 25 October 1854, during the late stages of the Battle of Balaclava, Lord Raglan, who was located far from the front, issued an unclear order to advance to the commander on the ground, Lord Lucan. The order was to be followed with immediate effect. The aide-de-camp who conveyed the order, Captain Nolan, had no respect for Lucan's abilities. Consequently, Nolan and Lucan failed to clarify between them the true meaning of Raglan's imprecise order. Lucan in turn conveyed the order to the Commander of the Light Brigade, Lord Cardigan. Lucan and Cardigan "detested"[25] each other, as a result of which final discussion of the order was minimal. The stage was set for the Light Brigade's charge into the teeth of the Russian artillery, and the rest is history. Figes reported 113 fatalities and 134 wounded out of 661 horsemen.[26]

The question here is, what possessed each of the cavalrymen to charge headlong into relentless Russian fire with such obviously poor odds on their side? Surely, fear, a primary emotion, would have impelled at least some of them to gallop away from the field of conflict instead. Yet, no such deserters were reported. This seems strange given the account of action selection given in chapter 8. On the face of it, the stochastic nature of the action selection mechanism of the brain would suggest that some of the Light Brigade should have deserted.

But they did not. Members of the Light Brigade almost certainly enjoyed the cut and thrust of warfare, the thrilling rage of conflict. On that analysis, then, the great majority of the Brigade's horsemen would have been furiously excited about the charge rather than fearful of the consequences. Armies are also famous for conditioning, or disciplining, their members to obey orders in all circumstances. As Cecil Woodham Smith put it:[27]

> Nevertheless, the Brigade made a brave show as they trotted across the short turf. They were the finest light horsemen in Europe, drilled and disciplined to perfection, bold by nature, filled with British self-confidence, burning to show the "damned Heavies" [the Heavy Brigade] what the Light Brigade could do.

Other primary emotional drivers were at work too. For a start, there was the fear that one might be shot for desertion. There was the further fear that, even if not shot, one would be deemed a coward amongst one's social circle and be shunned accordingly. The social emotion of shame would also play a part.

The answer to the question 'Why did each of the horsemen ride towards quite likely death?', is that selection and training and, in particular, timing minimised the chances of any of them thinking about the alternative course of fleeing. The last thing that their commander would have wanted was for any of them to have the chance to reflect on the forthcoming charge. So, this sounds like an instance of decision-making based on Type 1 processing, i.e., without conscious input, with a pinch of habitual learning thrown in too.

Compare this to the predicament of soldiers on both sides in the First World War waiting for hours, or even days, ahead of launching an attack from the trenches. Time and again during the war, the junior ranks were ordered to clamber out of their (relatively) safe trenches and charge the machine guns being fired by the enemy a mere few hundred yards away. Time and again, many of them obeyed and were duly shot by the enemy. These troops were different people from the horsemen of the Light Brigade. Many had volunteered, it is true, but without any sense of the horrors to come. The UK had not been in a war since the Boer War at the turn of the century. The Boer War was small by comparison with World War One. There was antipathy towards the foe. But, amongst some of the troops, this was leavened by some fellow-feeling for the suffering in the German trenches. Most importantly, there was the element of time for reflection. This was absent ahead of the Charge of the Light Brigade. But it was all too present in the long wait ahead of going 'over the top' in the First World War. This was the phrase that referred to the scramble up and over the side of the trenches following the whistle to attack.

One of the most famous encounters of the First World War was the Battle of the Somme. The battle unfolded over a period of months in the second half of 1916. It was spectacularly bloody with around one million killed and wounded. Here, in a letter published in Richard van Emden's *The Somme*, are some thoughts on one of the offensives in the battle, the Battle of Ginchy, on 9 September.[28] The writer of the letter was Second

Lieutenant Arthur Young of the 7th Royal Irish Fusiliers, writing to his aunt.

> It was about four o'clock in the afternoon when we first learned that we should have to take part in the attack on Ginchy. ... Well, even at the risk of making you ashamed of me, I will tell you the whole truth and confess that my heart sank within me when I heard the news. I had been over the top once already that week, and knew what it was to see men dropping dead all around me, to see men blown to bits, to see men writhing in pain, to see men running round and round gibbering, raving mad. Can you wonder therefore that I felt a sort of sickening dread of the horrors which I knew we should all have to go through? Frankly, I was dismayed. But, Auntie, I know you will think the more of me when I tell you, on my conscience, that I went into action that afternoon, not with any hope of glory, but with the absolute certainty of death.

Not far away from Ginchy and only 11 days later, Lieutenant Edward Tennant of the 4th Grenadier Guards wrote to his mother on the eve of another attack:[29]

> Tonight we go up to the last trenches we were in, and tomorrow we go over the top. ... I feel rather like saying, 'If it be possible let this cup pass from me', but the triumphant finish 'nevertheless not what I will but what Thou willest' steels my heart and sends me into this battle with a heart of triple bronze.

In these two cases, we have three warring emotions at a minimum: the primary emotion of fear (of the machine guns) and the social emotions of brotherhood (with one's comrades) and self-respect (or pride). Lieutenant Tennant clearly also has his religion as support. The real problem for Arthur Young, Edward Tennant and many other men was that their emotional conflict played out over days or weeks or months. It involved persistent, never-ending courage and strength of will. It carried none of the urgency of the Charge of the Light Brigade or of a rat suddenly encountering a cat. There was instead plenty of time for reflection, certainly enough time to wash away any feelings of patriotic *élan*. Can the analysis of the brain's workings as described by VBDM account for the determination shown by the junior ranks in World War One? Would not the primary emotion of fear of a brutal death impel most of the troops to flee in this case? Might this be the sort of occasion where consciousness is needed to generate the requisite behaviour?

Without an enabler such as consciousness, the stronger primary emotion of fear would likely overwhelm the more evolutionarily recent social emotions of companionship and self-respect. Without the persistence of decision-making over time, rather than just at a moment in time, the VBDM that operated initially to prefer companionship and self-respect over fear might well reverse. Companionship and self-respect might win the neuronal network conflict one moment. An hour later, with small changes to the neuronal environment, it might be superseded by its previously vanquished emotional rival, fear. The elegance and potential of the entire system would be undermined by the fact that its motivating forces could blow hither and thither. Troops might indicate they were ready for the fight one moment and flee the next, all in effectively random fashion. The whole enterprise would be hopeless. In this case, it is plausible to conclude that consciousness was the biasing input to the decision-making process that kept most of the troops in line most of the time.

But it remains the case that the decision-making process is stochastic. That means that the biasing input of consciousness was not flawless. Troops wavered and some did indeed give up. Personal crutches such as God or drink were needed by many. In addition, discipline was tough. Deserters were court-martialled and shot. So-called "whippers-in" also shot troops who went over the top and then turned tail.[30]

Decision-making with conscious input

As the brain grew in size in mankind's ancestors, and quite possibly in certain other non-human animals such as the great apes, elephants, whales and dolphins, it gave itself two problems. The first problem was that, although part of that growth in size accommodated the evolution of a raft of social emotions, how did the brain ensure that those potentially valuable social emotions were actually utilised? Human beings needed to behave in broadly predictable and prosocial ways to sustain their extremely complex social groupings. Human beings needed to be able to set themselves goals with a reasonable probability that they could see them through.

As noted earlier, a stochastic system is one that operates randomly, that gives rise to random outcomes. There are two difficulties with the purely stochastic approach to decision-making. First, the property of stochasticity does not, in and of itself, materially weaken or undo the demands of drivers as powerful as the primary emotions. On that basis, a social emotion such as sympathy would only make a rare appearance. Secondly, it would not be enough for a social emotion such as sympathy to be trotted out in a random fashion, as is implied by stochasticity. The subtleties, reciprocities and complexities of human society require its members to experience sympathy, for example, more than rarely and with some degree of predictability. Maybe this applies as well to other complex societies, such as those made up of apes or cetaceans. We have all come across people who are quick to anger or unpredictable. They tend not to fit into society as well as those who are more equable and reliable. A society where the majority displayed primary emotional behaviour on the whole and social emotional behaviour only randomly would be unstable.

The second problem had to do with cognition. Part of the brain's growth in size accommodated the evolution of Type 2 processing. But how did the brain ensure that this potentially valuable set of cognitive abilities was utilised in preference to automatic, relatively effortless Type 1 processing? The sheer power of our thought processes has given us human society as we know it today. One might argue that many facets of it are unattractive but no one can deny its complexity and the extent of its achievement. Type 2 processing is responsible for that but it often needed some help to prevail over Type 1 processing.

Conveniently, I propose, evolution gave birth to a new facility that improved the chances of decisions based on the social emotions and/or Type 2 processing triumphing over decisions based on the primary emotions and/or Type 1 processing, respectively. This was the biasing input that we call consciousness and that we experience as Self-as-subject. I propose that consciousness addresses the two problems cited above. Those hominids with such a facility would quickly out-compete those without.

A challenge lies in how to discuss consciousness without falling into the homunculus trap. Consciousness is a *deus ex machina* that exists to shade the odds in favour of more predictable, more thoughtful behaviour. But this god must be mindless; otherwise, it would be a homunculus. It must possess an indirect causality, not be directly causal. Even if it is mindless, this god cannot feel mindless. If it did, it would lose its efficacy. Let us get to know Self-as-subject better.

Chapter 15

Self

Self-as-subject and Self-as-object

In part 2, we introduced the idea that the brain possesses a biasing mechanism, consciousness, which promotes the social emotions and Type 2 processing, respectively. In part 3, we explored the opaque topic of consciousness in more detail. It is time to draw the threads together. The first step in this process is to challenge our intuition that 'I', or 'I myself', control my own body and brain. Almost all of our daily language presumes that this is how we work. It tends to be left to philosophers to lift a cautionary finger. One might say, I tried really hard to get the top grade in my mathematics paper but I failed. What does 'I' mean in that sentence? At a trivial level, it is helpful for one's interlocutor. It is better to say that than, referring to myself, Rupert tried really hard to get the top grade in his mathematics paper but he failed. The person to whom I was speaking would be confused, rightly. We all have a deep-set intuition that each of us possesses an 'I', but what exactly constitutes an 'I', or a self?

Self appears to be inextricably linked to consciousness. Indeed, self-awareness and consciousness are often thought to be the same. Yet, self and consciousness are different things. The term 'self-awareness', awareness of self, suggests that self is an object. Moreover, consciousness sounds like awareness (of that self). But, if there is an object and if there is awareness of it, then there must be a subject. Something has to be conscious or aware of self-as-object (Self-as-object). Practically speaking, the only candidate for that something is self-as-subject (Self-as-subject).

Antonio Damasio defines Self-as-object as:[1] "*... a dynamic collection of integrated neural processes, centered on the representation of the living body, that finds expression in a dynamic collection of integrated mental processes.*" Thus, Self-as-object is an explicitly physical entity, which is reflected in the term "integrated neural processes". These processes take the form of our perceptions, our memories and our thoughts. They arise from the activities of, and inputs to, our bodies and brains. If we were not embodied, our neurons would not be able to capture any information to process.

Damasio goes on to define Self-as-subject as follows: "The self-as-subject, as knower, as the 'I', is a more elusive presence, far less collected in mental or biological terms than the me [self-as-object], more dispersed, often dissolved in the stream of consciousness." We clearly need to examine this proposition further since it seems counter-intuitive. Most of us have a powerful sense of self, or at least we think we do. But impressions can be misleading.

Self, therefore, is not a single thing, even if it feels like it. The feeling of a unified self is misleading. There are two selfs, or selves: Self-as-subject and Self-as-object. Self-as-object is a collection of neuronal networks in the brain, Damasio's "integrated neural processes". One could also term it 'the unconscious mind'. Self-as-subject, on the other hand, has an elusive and slippery nature. It might be termed 'the conscious mind'. A clear understanding of the forms that Self-as-subject and Self-as-object each take and the way they develop and interact is fundamental to the formulation of a coherent theory of consciousness.

Self-as-object is easier to pin down than Self-as-subject. You are "nothing but a bundle or collection of different perceptions", to quote the great Enlightenment philosopher, David Hume. This is a quotation adduced by philosopher and writer Julian Baggini to explain so-called "bundle theory".[2] To put it simply, Self-as-object possesses the substance of self in the form of all our perceptions, thoughts and experiences; hence the term 'bundle'. These are expressed in countless neuronal networks in our brains. On the other hand, Self-as-subject has limited substance to speak of. Self-as-subject's perceptions, thoughts and experiences actually originate with Self-as-object. Self-as-subject might feel them but does not make them. Self-as-subject has a "sense of ownership" of their body.[3] You feel that you, not some collection of unconscious neuronal processes, see the birds and the leaves on the trees. You feel that you, not some bundle, has a stomach-ache or an ear-ache, for example. Self-as-subject also has a "sense of agency".[3] You feel that you, not some hidden force, are lifting your arm or kicking a football, for example. So, Self-as-subject clearly exists. The question is what form it takes.

Getting to know Self-as-subject better

The most important feature of Self-as-subject is that it should feel authoritative. I decide to do *x*, and then as a result I do *x*. It has to feel authoritative. It needs to bias our stochastic decision-making processes away from our primary emotions and in favour of steady prosociality. It needs to be able to promote Type 2 processing. It also has to achieve these tasks in an indirect way. We have established that, if Self-as-subject were directly causal, then it would be a homunculus. We cannot have that.

We all know what authority figures look and sound like. They are agents, to use a philosophical term. They are aware of themselves and their surroundings. They have and express views and, where appropriate, accompany those with actions. They have purpose, both generally and specifically in the form of discrete goals. Finally, an authority figure means something to us. Our parents constitute the first authority figures that most of us encounter. At that age, they mean the world to us. This distinguishes such authority figures from other agents. They have views and behaviours that we respect and take seriously. That is quite a combination of positive attributes; it would be worth possessing as well. What if we had a hybrid authority figure inside our heads to help us navigate the world? That would be welcome. We do of course, in the form of Self-as-subject. This feels like a homunculus, or little man or woman, sitting inside our heads (see figure 15.1). Note, though, that this is not an actual homunculus. We are lucky not to be zombies: they do not possess any such internalised authority figure.

Figure 15.1
Self-as-subject
as internalised
authority figure

This is part of the answer to Chalmers' "hard problem of consciousness". Consciousness needs to feel like something in order that it can influence our decision-making processes. If it did not feel like something, then it would be just another part of the unconscious mind participating in the winner-takes-all action selection process. So, it would be prey to the same problems as those already discussed. For example, the primacy, all other things being equal, of primary emotions and Type 1 processing would remain unchanged.

The reason that we feel that Self-as-subject must carry weight and authority is that we each have a sense of deciding what to do next, what actions to take. Generally, when we do something purposeful, we do not feel as if someone else has taken over; nor do we feel that the action in question has been taken at random. That very sense of deciding, of being an agent, is Self-as-subject at work within us. It feels as if Self-as-subject has all the attributes of the authority figures around us: agency, awareness, purpose and respect. Provided that I am mentally fit and healthy, I feel through Self-as-subject that I am doing stuff, that my body is obeying what I command it to do. I am aware of my body and of everything that my five senses perceive and, indeed, of some of the goings-on in my brain.

Conversely, if Self-as-subject had no weight or authority, then we would drift in an aimless fashion. It would have no impact on our decision-making. We would still be an agent but one with no purpose. We would possess consciousness but, *in extremis*, would do nothing with it. People with severe depression, for example, fall into this category. Consciousness did not evolve so that we could wander around without purpose. It evolved as a biasing input and to that end needs to feel weighty and authoritative.

Moreover, if Self-as-subject did not have access to the processed outputs from our exteroceptive (sight, smell, etc.), proprioceptive (body position) and interoceptive (body state) senses, then it would be unable to participate in our decision-making processes in the right way and at the right time. It would be random, which is precisely what the evolution of Self-as-subject is selecting against.

What form, then, does Self-as-subject take? If it is not to be a homunculus but nonetheless has the attributes described above, then it needs to be instantiated in the brain as a neuronal network. Moreover, it will need to take the form of a network highly interconnected with the rest of the brain. In short, whilst Self-as-subject may appear "elusive", it is nonetheless present in the brain in the form of a neuronal network.

Important Self-as-subject network connections with the rest of the brain

Let us explore the location of the Self-as-subject network. Joseph LeDoux notes that[4] "Brain areas that have been implicated in self-processing" include "ventromedial prefrontal cortex ... frontal pole ... anterior cingulate cortex ... dorsomedial prefrontal cortex ... posterior cingulate cortex and precuneus." The exact location of each of these areas is unimportant for our purposes, although some of them can be picked out in figures 5.2 and 6.1.

It would be premature to conclude that the Self-as-subject network is, therefore, instantiated in the regions of the brain listed above. But the list reinforces the point that the Self-as-subject network is likely to be extensive. Moreover, it will be highly connected with other parts of the brain too. It must, for example, connect with the brain's full suite of memory capabilities, especially episodic memory. This is the type of memory that recalls personal experiences of things rather than just the things themselves. In this way, Self-as-subject can be triggered to participate in the decision-making process when a current situation is similar to a previously experienced situation.

In humans, another important part of the network's connectivity is with the brain's language areas. Access to the language output from our brains enables Self-as-subject to communicate, to give effect to its authoritativeness. This is why we experience our inner voice, what Jackendoff called the "talking voice in the head, the so-called stream of consciousness".[5] A human's inner voice gives a massive step-up in both clarity and also impact and, hence, authority to Self-as-subject. For this reason alone, consciousness in humans is of an order of magnitude more efficacious than in any conscious animal without language. For the avoidance of doubt, the language used by Self-as-subject is provided by Self-as-object. The words are merely routed through Self-as-subject. The effect of this routing is to give the words greater salience when they reappear in the brain's decision-making region. At the risk of labouring the point to death, Self-as-subject does not generate language itself, nor is it directly causative.

Some of the connections of the Self-as-subject neuronal network are more important than others. Most obviously, the Self-as-subject network must be highly interconnected with, and receptive to, the networks that instantiate the social emotions. One of Self-as-subject's two main jobs is to increase the chances of social emotions prevailing in decision-making. As explained in chapter 6, we do not know in detail how the social emotions are represented in neuronal form. But the world of psychology has made significant progress in establishing how they develop in humans in childhood. We look at this in the next section.

Another important connection for Self-as-subject has to do with causality. Self-as-subject needs to be experienced as an agent. It needs to feel as if it is causing things to happen, even though that causation is in practice purely indirect. Manos Tsakiris and Simone Schütz-Bosbach, both professors of psychology, and Shaun Gallagher, a professor of philosophy, have pointed to research evidencing "the neural correlates of the sense of agency"[6] in the motor regions of the brain. Through Self-as-subject, we need to feel, where appropriate, that we have caused whatever action has been executed. In the first instance, this is underpinned by the brain's comparison of its perception of the action taken with a copy of the instructions for the action originally sent to the brain's motor regions. There is evidence that such comparisons occur towards the front of the right parietal cortex and, in front of that, in the insular cortex.[6] This area would be connected to the Self-as-subject network in order for that sense of agency to be experienced.

Large-scale neuronal networks such as that conceptualised in the form of the Self-as-subject network are a relatively new idea in the world of neuroscience. Amongst these networks are the "frontoparietal control network", the "default network" and the "dorsal attention network".[7] Research is taking place to establish both the functionality of these networks and their connectivity with each other and elsewhere in the brain. The Self-as-subject network differs in one critical sense from such networks. Its functionality is minimal even though its connectivity is substantial. The Self-as-subject network does not possess direct causality. If it did, it would take the form of a homunculus, which is off-limits.

Disentangling the elements of the Self-as-subject network from other large-scale neuronal networks such as the frontoparietal control network is currently beyond our capabilities. We can detect correlation of neuronal activity with behaviour. We are, however, unable to identify which parts of such neuronal activity are directly causative and which parts are indirectly causative. The latter would reflect Self-as-subject activity.

Development of the Self-as-subject network in infants

So-called object relations theory aims to explain how the infant develops distinct mental representations of itself and of others. A newborn's brain contains a huge amount of processing potential but limited content. Indeed, in *Affect Regulation and the Origin of the Self*, Dr. Allan Schore notes that:[8,9] "the electroencephalogram of the human in the first month of life if derived from the scalp of an adult, would be considered 'sufficiently abnormal to indicate imminent demise'."

In order to function in the world, the infant's brain must develop its own Self-as-subject neuronal network. The physical interactions between the infant and the primary caregivers are fundamental to the development of this network.

Schore and others, such as psychoanalytic psychotherapist Sue Gerhardt, identify the first three years of life as the critical formative period for human beings.[10,11,12] Gerhardt observes that "The period from conception through the first two years—roughly speaking, the first 1000 days—is uniquely significant, because this is when the nervous system itself is being established and shaped by experience." This does not mean that an individual's personality is fixed in stone thereafter. But it does mean that this first period of three years or so is uniquely important relative to the rest of that individual's life, barring significant subsequent trauma, for example.

The first significant phase of this three-year period constitutes months one to nine. Schore describes how the primary interaction between infant and mother is "sustained mutual gaze"[13] (see figure 15.2). At a psychological level, this type of interaction energises the infant. This is not to say that the child has no motivation on its own to explore and develop, but interactions with its mother significantly invigorate it. At a neurological level, such interactions are essential.[14] The infant's brain actually changes and develops as a result of the intense intimacy of the gazing. Gaze is not the only pertinent interaction. Language, particularly of the soothing kind, and touch are equally important to the infant in beginning to understand the world and hence his or her emotional interaction with the world.

Despite this progress, the infant remains unable to regulate its own emotions: the Self-as-subject network has not yet developed. The mother has to fulfil this role on behalf of the infant, which she does through gazing. In addition, she may comfort and hold the infant if it is in distress or in a rage, and perform other self-calming actions that Self-as-subject will take over in due course.

Figure 15.2
The development of the Self-as-subject network (SaSN) in infants

The second significant phase of this three-year period takes place over months 10–18.[15] The mother makes two important changes to the routine. First, the mutual gazing lessens in frequency. Secondly, in the latter part of this period, the mother introduces the idea of disapproval to the infant. This clearly creates tension between the two.[16,17] Disapproval of poor behaviour is telegraphed to the infant by the mother through looks of embarrassment, distaste, disgust or even withdrawal of gaze. They arouse feelings of shame in the infant. Neurologically speaking, this shift in the mother's approach helps lead to the "maturation of the orbitofrontal cortex".[18] So, these interactive behaviours are continuing to cause real changes to the wiring of the infant's brain.

Shame may not be the only motivating force in the infant's brain. Other forces could be discomfort, confusion and disappointment. Infants and small children will look at their parents to find a cue in what they perceive in order to mimic that feeling or disregard it. Parents will in turn amend their expressions to provide a relevant cue. The process is iterative.

I believe that the infant's brain will have developed a large part of its Self-as-subject neuronal network by this stage. There is more to come. Schore describes how the other caregiver, typically the father in his account, plays a greater role than previously in the development of the infant's brain from the age of 18 months.[19] This is the start of the third significant phase of the infant's three-year period of personality formation. Phase three ushers in self-regulation with words. Once again, the infant's brain actually changes to achieve this. Schore locates this maturation in the dorsolateral cortex in the left hemisphere.[20]

I propose, therefore, that the foundations for the Self-as-subject neuronal network take up to three years to form within the infant's brain. Its formation consists of millions of new synapses, relying on the principle of synaptic plasticity described in chapter 5.

With the foundations laid, the Self-as-subject network still needs more time to cohere. In *The Philosophical Baby*, Alison Gopnik, a professor of psychology with an affiliate professorship in philosophy, describes a telling experiment run by Danny Povinelli.[21] The experiment tested the different reactions of three-, four- and five-year-olds to finding a sticker surreptitiously placed on their foreheads. In each case, the child was video-taped and then the video was immediately played back to the child. Gopnik reports the results as follows.

> Tellingly, the three-year-olds ... referred to the child on tape by using their own names, while the fours said that the child on the tape was "me". At three Johnny would say, "Look, Johnny has a sticker on his head," and make no attempt to touch his own head. At four he would say, "Look, I have a sticker on my head," and immediately reach to take it off.

Five-year-olds were even quicker to react. As Gopnik observed, "children got much better at this kind of self-control between three and five."[22]

It does not stop there. Gopnik reports that four- and five-year-olds "deny experiencing visual imagery or inner speech".[23] They do, however, talk to themselves out loud. Similarly, whilst they exhibit episodic memory, Gopnik explains that they cannot string their memories of past events together.[24] Gopnik concludes that it is not until they are six that they appear to possess these faculties.[25] The Self-as-subject neuronal network is, therefore, built, I believe, in stages during childhood. It is the physical manifestation of the neuropsychological developments described above.

One final note should be added. The process described above has been going on now for hundreds of thousands, if not millions, of years. After all, there have been plenty of types of human being, modern and archaic, over the last few millions of years. Many, if not all of these, are likely to have invested heavily in child-rearing as the length of childhood grew during this period in parallel with increasing post-natal brain development.[26] The result of this is to predispose human brains in infancy to develop Self-as-subject networks even under the most difficult of early upbringings.

Revisiting value-based decision-making (VBDM)

If the above account of the Self-as-subject neuronal network being instantiated in the brain is to hold water, then the Self-as-subject that emerges must be able to be accommodated within the VBDM framework. In humans, for example, unlike any in other conscious animal, language has significantly changed how the brain works. The human Self-as-subject network performs much of its work through the device of language.

Self-as-subject's connection with the language areas of the brain is intricate. Recalling Jackendoff, the unconscious mind is responsible for producing the words of our inner voice. But these words are first generated in mentalese, the language or code of the unconscious mind described in chapter 10. Maybe the neuronal network underlying a certain social emotion is triggered in a particular situation. How does that social emotion make itself heard over the din of the primary emotions? How does the brain plump for sympathy, for example, over the hot and urgent dictates of anger? Clearly, we feel sympathetic. That is to say, Self-as-subject experiences the underlying emotion of sympathy. But Self-as-subject might well also need to be heard in order to promote sympathy in the decision-making process. In humans, that is partly achieved through the medium of language.

The axons from the social emotion (sympathy) neuronal network need to connect with the dendrites of the language areas of the brain in order to translate its mentalese-expressed impulse into English or Spanish or whatever language suits. Using the internalisation device that is your inner voice, Self-as-subject then loops the relevant social emotion, expressed verbally, back into your decision-making.

It is not, however, a one-way street. A conflicting primary emotion (anger) might fight back. Again, the emotion of anger will be felt. But what is sauce for the goose is sauce for the gander. Anger is quite likely to be verbalised too. Rendered in neuronal network terms, the axons from the primary emotion (anger) neuronal network connect with the dendrites of the language areas of the brain. Consequently, a conflicting emotion is translated into your mother tongue. It too loops through the Self-as-subject that is you and back into your decision-making by way of your inner voice.

Consider: these verbalised outputs are exactly what you experience when you are agonising with yourself over a decision. You might experience, for example, a prosocial emotion such as remorse, were you to take a certain bad action. The feeling of remorse will likely be accompanied by appropriate words in your head. A second later, a more urgent, more arresting emotion such as anger with an accompanying message might come through, trying to knock the prosocial emotion out of play. Agonising decisions might go back and forward like this for hours, or even days.

But then you decide, one way or the other. And this is the critical step in the argument: 'decide' means purely that the winner-takes-all process in the brain causes the brain to go one way or the other. Self-as-subject does not decide. Self-as-subject is not a homunculus. Self-as-subject is only a neuronal network. I suggest that it is heavily connected to the social emotions, and less so to the primary emotions. By dint of such connections, the social emotion neuronal network might be enabled to compete more vigorously in the winner-takes-all process. Indeed, in a battle between anger and sympathy, for example, you might tell yourself to rein in your anger as well as showing sympathy.

The development of Self-as-subject gives rise to the further benefit of promoting Type 2 processing relative to Type 1 processing. It does not do this directly. It does not perform Type 2 processing itself. Rather, it acts through the decision-making process by biasing the brain towards persistence despite the extra energy consumption entailed by Type 2 processing. The temptation to accept the easy answer, often inspired by one of the primary emotions, is shouldered to one side during the winner-takes-all process. Typically, a social emotion lies behind this. Revising hard for exams might ultimately be driven by pride or a desire to please or competition. Once again, though, it is not guaranteed that Type 2 processing will take place. It merely might.

I say 'might' because the impact that Self-as-subject has on decision-making in any given situation depends upon a number of factors. Energy is one such factor. Daniel Kahneman relates the disturbing story of a research experiment involving eight parole judges in Israel reviewing parole applications.[27] The judges were fed three times during the day. After each meal, the rate of parole application approval shot up to as high as 65%. It then drifted down "steadily, to about zero just before the [next] meal". It is just easier to say no if one is running out of energy. If one is feeling tired, one reverts to Type 1 processing, which is automatic and features no deliberation or executive control.

Another factor is one's strength of will. This is a measure of the impactfulness of one's own Self-as-subject. This is an immensely important point. It means that each human possesses his or her own strength, or intensity, of consciousness. Like the varying size of the muscles in different people's arms or legs, or people's differing degrees of hand–eye coordination, the strength of each person's consciousness will vary. It certainly varies depending upon how energetic one is feeling. But, just as someone might possess bigger or smaller muscles than other people, so do we all possess greater or lesser degrees of potential strength of consciousness. By strength, I mean Self-as-subject's ability to promote the social emotions or to persist with Type 2 processing.

It might be objected that dictators such as Hitler, Stalin and Mao were all strong-willed yet not remotely prosocial. Regrettably, however, not all the so-called social emotions are in all cases prosocial. Pride, if taken to extremes, is a good example. Pride can become overweening, leading to uncongeniality and bad behaviour. *Schadenfreude* (taking pleasure in others' misfortune) is another good example. Dictators exhibit in an extreme sense a number of highly unattractive social emotions.

A last word

There are immense benefits from looking at consciousness through the lens of the Self-as-subject neuronal network. Obviously, I believe it to be the right lens. More importantly, however, it roots the explanation of consciousness in firm biological foundations. Most scientists believe that consciousness is capable of an explanation based on biology, specifically the biology of the brain. The hard work arises when it comes to explaining how such biology works in practice. Like Rolls, I believe that the answer to this challenge lies in neuronal networks.

There is an incidental benefit, which goes to the point of this book. Whilst artificial neural networks are still some way away from mimicking neuronal networks, there are nonetheless strong parallels between the two. If we are to instantiate consciousness in a robot, in AI, then the solution will be by way of replicating a neuronal network-based explanation of consciousness *in silico*. Let us first, though, explore the remarkable discovery of a special type of neuron in the brain, the mirror neuron.

Chapter 16

Mirror Neurons

Introduction: a special type of neuron

In 1992, Giuseppe di Pellegrino, a professor of psychology, *et al.* published a paper reporting on the discovery of a type of neuron in monkeys' brains with mirroring properties.[1] The findings were surprising; note the words *also* and *observes* in the last sentence of the following quotation.

> Neurons of the rostral part of inferior premotor cortex of the monkey discharge [fire] during goal-directed hand movements such as grasping, holding, and tearing. We report here that many of these neurons become active also when the monkey observes specific, meaningful hand movements performed by the experimenters.

You would expect the neurons in your own motor cortex to fire when you yourself are moving your hand. What was deeply surprising about the discovery was that it suggested that some of your motor neurons would also fire when you watched someone else moving their hand. Mirror neurons had been spotted for the first time.

As Giacomo Rizzolatti, a professor of physiology with a special interest in neurophysiology, and Corrado Sinigaglia, a professor of philosophy of science, point out in their book, *Mirrors in the Brain,* it had been known for a while that certain neurons in the premotor cortex possessed so-called "visuo-motor properties".[2] In other words, despite being in the motor regions of the brain, such neurons have a "role in the process of transforming the visual information regarding an object into the appropriate motor acts".[2] So, they help the monkey (in this case) grasp an object. But mirror neurons fired not just when the monkey itself grasped an object, but also when the monkey saw a researcher (or another monkey) grasping that object. The monkey's mirror neurons only fired when an action was perceived. They did not fire merely at the sight of the object in question. Human mirror neurons seem to be less limited in scope.[3]

The discovery that mirror neurons fired explicitly in response to the behaviour of others was clearly intriguing. Rizzolatti and Sinigaglia cite a number of possible functions of mirror neurons. These embrace "*action understanding*", imitation, empathy and possibly aspects of language.[4,5,6,7] They identify the "*primary* role" of mirror neurons, though, as action understanding.[8] Action understanding means that the observer understands not just what the observee is doing, but also why. Mirror neurons enable the observer to understand the observee's intention. Specifically, they argue that the replication in the observer's premotor cortex of premotor cortex neuronal activity in the observee's brain "generates a *basic motor knowledge* of the meaning of the acts coded by the various neurons".[9] As a result, the observer "immediately *perceives the meaning* of these 'motor events' and *interprets them* in terms of an *intentional act*".[10]

Mirror neurons have sparked considerable debate, with both supporters and detractors. The argument is not about whether they exist. Rather, there is considerable disagreement about their purpose. Unhelpful to the cause of mirror neuron supporters is the fact that it remains difficult to make single-neuron recordings deep in the human brain. There is the risk of significant damage. So, witnessing at a neuronal level what the human brain does with the information processed by its mirror neurons so far eludes us.

The mirror neuron naysayers, however, must contend with the fact that it now appears that the mirror neuron system is extensive. Moreover, there are many mirror neurons: "about 20 percent of the cells in area F5 of the macaque brain are mirror neurons."[11] As such, they consume energy. When part of the body consumes energy on a regular basis, there tends to be a good reason. It is too exhausting for an organism to expend energy wastefully.

How mirror neurons work

Let us start with the original mirror neurons discovered in 1992, as described by Rizzolatti and Sinigaglia. To recapitulate, these mirror neurons were located in the premotor ventral cortex of monkeys. The motor regions of the brain are divided into a number of smaller areas, each of which have subtly differing functions. In broad terms, they vary as to the extent that they interact, for example, with sensory regions of the brain, with planning circuits and with muscle movement circuitry. Perhaps most importantly, part of the motor cortex needs to combine visual inputs with movement. There is no point in the monkey reaching for a banana if the monkey's brain has not integrated the precise location of the banana with the hand trying to grasp it.

The monkey's mirror neurons only fired at the sight of motor acts performed by others in relation to objects.[12] In a further twist, the mirror neurons did not fire if the motor act was faked.[12] Each mirror neuron differed from the others. Specific motor acts triggered specific mirror neurons.[12] They were not multi-purpose. Approximately 70% of the mirror neurons were "broadly congruent" and fired at types of acts.[13] Other mirror neurons, "strictly congruent" ones, fired only at specific acts.[14]

Marco Iacoboni, a professor of psychiatry and biobehavioural sciences, points to yet another type of mirror neuron.[15,1] This is the "'logically related' mirror neuron[, which], for instance, fires at the sight of food being placed on the table and also while the monkey grasps the piece of food and brings it to the mouth". In this way, mirror neurons may be able to identify the intention behind, in this case, putting the food on the table.

The next step was to discover whether humans possessed a mirror neuron system as well. The difficulty with humans, as pointed out earlier, is that it is risky to insert electrodes into the human brain. Using other, non-invasive techniques, such as transcranial magnetic stimulation (TMS) and functional magnetic resonance imaging (fMRI), the answer quickly emerged that humans appeared to have mirror neurons too. The human mirror neuron system was, in some regards, more sophisticated than that of monkeys. Rizzolatti and Sinigaglia reported that the human mirror neuron system was also active when the observee performed an action without any object being involved.[16]

The discussion above extends just to the nexus of vision and action. In *The Empathic Brain*, Christian Keysers, a professor of social neuroscience, described how mirror neuron activity can be initiated by the auditory system as well.[17] Rather than certain neurons in the premotor cortex firing at the sight of observed actions, they fired at the sound of observed actions. This ability was detected in both monkeys and humans.

Keysers also described how the mirror neuron system seemed to extend to the perception of, or empathy with, emotions in others.[18,19] A new part of the brain was involved in delivering this capability. This was the (anterior) insula, in the folds of the neocortex.[20] It turned out that the mirror neuron system could be activated in other ways as well. For example, reading about something disgusting could activate the system too.[21,22]

The mirror neuron system is also activated by the perception of people interacting with each other. In *Mirroring People*, Iacoboni refers to an experiment that recorded the mirror neuron activity of subjects observing a number of different video clips.[23] "[H]alf depict[ed] communal sharing social relations, the other half depict[ed] authority ranking social relations." The videos, especially those portraying "communal sharing social relations", strongly activated the subjects' mirror neurons.

A meta-analysis of 125 human fMRI studies concluded that the human mirror neuron system is extensive, being situated across various regions of the brain.[24] It consists of the classic mirror neuron regions, primarily the premotor cortices, as well as additional, complementary areas.

One problem remained. fMRI allows one to infer that humans possess mirror neurons. But that is not the same as seeing them directly. In 2010, Roy Mukamel, a professor in psychological sciences, had a fortunate breakthrough. One of his colleagues, Itzhak Fried, a professor of neurosurgery, psychiatry and biobehavioural sciences, was working with patients to treat cases of epilepsy. The treatment involved implanting electrodes into the affected region of the brain for diagnostic purposes. So, the patients' brains were already wide open, so to speak. The opportunity presented itself, with prior approval by the patients, to make recordings of single neurons as the patient observed others grasping objects or making facial expressions. The results confirmed directly that humans possessed mirror neurons.[25]

Even the origins of mirror neurons have been the subject of considerable debate. The three main contestants are the so-called associative, genetic and epigenetic accounts of their origins. There appears to be a majority in favour of the idea that the mirror neuron system is largely shaped in infancy and early childhood. I shall return to mirror neuron activity in infants and young children in the next chapter, "The Mirrored Homunculus".

What purpose do mirror neurons serve?

In 2017, Rizzolatti and Leonardo Fogassi published a paper reviewing the full range of possible functions of the mirror neuron system.[26] These constituted the understanding of others' emotions, the understanding of others' goals and intentions during action, imitation and the development of language. We shall briefly explore each of these in turn, starting with imitation. We shall, however, not consider the possible involvement of the mirror neuron system in the development of language; it is not pertinent to the current discussion.

The utility of mirror neurons in imitation is perhaps the least controversial of possible mirror neuron functions, although there is no agreement on whether it is the most important function of mirror neurons. Rizzolatti and Fogassi, like others, divide imitation into two varieties, "replication" and "learning by imitation".[27] Replication might also be called mimicry; it tends to be used in connection with simple actions. Learning by imitation is used for more complex, novel actions. They connect mirror neurons to both types of imitation. Cecilia Heyes, a professor of psychology with a particular interest in mirror neurons, and

Professor Greg Hickok, one of the bigger critics of most mirror neuron theorising, both broadly agree that mirror neurons play an important role in imitation.[28,29,30] Hickok adds the interesting and important point that the better the brain, the more benefit it can derive from mirror neuron functionality. Humans, therefore, get more out of mirror neurons than monkeys.[31]

Rizzolatti and Fogassi also touch on neo-natal imitation,[32] which is the ability of newborns to imitate facial expressions such as sticking out their tongue. This ability "disappears in humans and chimpanzees at about 2–3 months of age". They conclude that the ability relies on a relatively small mirror neuron system that is "hardwired" at birth, a sort of mirror neuron starter-kit.

Now we move to more controversial territory. Rizzolatti and Fogassi support the idea that mirror neurons play an important role in understanding the emotions of others, as described above. On this, Hickok disagrees.[33] The most controversial potential purpose of mirror neurons, though, is action understanding. Iacoboni *et al.*, including Rizzolatti, have pointed out that mirror neurons seem to be differentiated as between action-perception and intention-perception.[34] In their experiment, some mirror neurons were activated in response to a perceived action, or motor act. Other mirror neurons were activated in response to the intention behind the action, i.e., why it happened. Their research found that the context in which the action was carried out was instrumental in activating intention-related mirror neurons. The ability of the mirror neuron system to perceive intention is facilitated by the logically related mirror neurons mentioned earlier.

The proposal for action understanding, then, includes the proposition that some mirror neurons enable an observer to internalise the perceptions of an observee's actions and the intentions lying behind them. This internalisation in turn enables the observer to understand what an observee is doing and why. This is sometimes called simulation theory, in that we are able to simulate in our minds an observee's intentions.[35]

In this light, Iacoboni *et al.* divide intention into "immediate, stimulus-linked, 'intention' or goal" and "global intention".[34] The former refers to Mary's goal in reaching for an apple: Mary's goal or immediate intention is to grasp it. The latter refers to Mary's broader or global intention thereafter, in this case perhaps to eat it. They suggest that logically related neurons are responsible for divining global intention behind others' actions. Rizzolatti and Sinigaglia distinguish between such a global intention and an even deeper intention. The even deeper intention might be that, say, she wanted to stop her younger sister having any of the apple. As they observe,[36] "Understanding the reasons behind an agent's motor intention requires additional inferential processes",[37,38,39] which we as yet do not understand.

Hickok disagrees with the proposed action understanding function of mirror neurons and cites a number of "anomalies" in the evidence to make his point.[40] By way of example, he notes that we can understand actions that we cannot perform: "When my cat purrs she is socially receptive; when she hisses, I need to back off."[41] He says that we understand such actions even though we do not do them, and so do not have a mirror version of them in the motor regions of our brains. A counter-argument would simply be that the brain relies on primary emotional circuitry, in this case, fear, to generate the withdrawal behaviour. Our reaction is instinctive and, so, no mirror processing is needed.

More generally it should be noted that, like any other set of neurons operating in a network, mirror neurons cannot guarantee the outcome. When they simulate, they are effectively pointing to the most likely interpretation of the action. This turns out to be an important part of the process that leads to the development of consciousness.

What happens to the information processed by the mirror neuron system?

A further interesting question is what the brain then does with the information that it has mirrored. To put the question another way, what other neuronal activity is inspired by the firing of the relevant mirror neurons? In order for mirror neurons to facilitate imitation, their activity must lead to changes in the brain in order for the imitation to take place on the relevant subsequent occasion. The firing of the mirror neuron system needs to precipitate the formation of a permanent neuronal network dedicated to the action or series of actions that the observer is observing. Otherwise, how will the brain be able to imitate the relevant action in the future? This is termed imitative learning.

The formation of this new network will be based upon the principles of neuroplasticity introduced in chapter 5. We learn to play tennis by observing the coach, and then having a go ourselves. There would be no point in the mirror neuron system firing in our brains as we observed the coach playing a topspin forehand, and then doing nothing else. The lesson would not stick. The firing of the relevant mirror neurons needs to be a precursor to the instantiation in our brain of a permanent neuronal network. That network needs to be associated with an action, e.g., to play the topspin forehand in certain circumstances. It also needs to be associated with a goal, e.g., to hit a forehand winner with said topspin. The action becomes increasingly automatic in the way that it only can if a permanent topspin forehand neuronal network is formed, bit by bit, in our brains. We call this muscle memory but really it takes place in the brain.

Giovanni Buccino *et al.* investigated imitative learning using guitar playing as their research environment.[42] They asked their research subjects, who were new to guitar playing, to participate in a guitar lesson whilst their brains were monitored using fMRI. "Musically naïve participants were scanned during four events: (1) observation of guitar chords played by a guitarist, (2) a pause following model observation, (3) execution of the observed chords, and (4) rest." Perhaps most interesting was the discovery that part of the prefrontal lobe became engaged during phase two of the experiment. This was the phase where participants were digesting what they had observed before going on to play themselves. As Rizzolatti observed of the experiment,[43] Buccino *et al.*:

> proposed a two-step processing in imitation learning: first, "mirror" activation of motor act representations in the parietal and frontal lobe; second, the recombination, thanks to prefrontal lobe (area 46), of these motor acts, so [as] to fit the observed model. The same authors carried out a subsequent fMRI study in expert and naïve guitarists. This new study confirmed the previous data. In particular, the data showed again the fundamental role of area 46 in combining different motor acts in a new motor pattern.

The fact that the mirror neuron system must lead to change in the brain is very important, as we shall see in the next chapter. Chapter 17 will concern itself with a more speculative theory of a possible function of mirror neurons. It constitutes the core of my theory as to how consciousness is instantiated in the human brain.

Chapter 17

The Mirrored Homunculus

The mirrored homunculus theory in a snapshot

I propose that each of us develops consciousness in our formative years. I suggest that, in a child's formative years, its brain mirrors the self that it perceives in its parents or caregivers. In other words, the child perceives its mother, for example, as being an agent, being herself. Just as the child's mirror neurons mirror its mother's motor neurons lifting her hand up or smiling, for example, so do the child's mirror neurons mirror the self that they perceive in the mother. This process of self mirroring clearly does not happen in one go. The self that is mirrored is built up neuron by neuron by way of imitative learning.

The early brain effectively perceives an agent or homunculus in others by way of its mirror neurons. It thereby mirrors that, in neuronal network form, in itself—the mirrored homunculus. The mirrored homunculus at this stage is no more than a neuronal network with extensive connectivity to the rest of the brain. As I already pointed out in chapter 8 and elsewhere, there is no such thing as a real homunculus. But the infant brain does not know that. Its mirror neurons merely mirror what they perceive in others, a self-as-subject neuronal network. Having been mirrored, this self becomes instantiated in the child's brain in the form of an extensive neuronal network, the child's own Self-as-subject network.

Self-as-object interacts extensively with Self-as-subject as a result of the sheer reach of the Self-as-subject network across the brain. The interesting question is how Self-as-object perceives Self-as-subject as a result of this interaction. It mistakenly perceives Self-as-subject as an agent, which is an illusion. This illusion is consciousness, as we shall see in further detail in the next section. The Self-as-subject neuronal network is not consciousness. Rather, it underpins consciousness. It is Self-as-object's illusory perception of Self-as-subject that is consciousness. This is the point at which the mirrored homunculus achieves its full strength. It turns out that consciousness is a fortunate accident, an evolutionary by-product of mirror neurons that persisted because it had value to its possessor.

The mirrored homunculus develops over the first several years of a child's life. It is instantiated in the form of one of the most extensive neuronal networks in the brain. This instantiation takes place by way of synaptic plasticity, a process initiated in this case by the mirror neuron system. The Self-as-subject network comes into being as an overlay on top of the brain's entire, pre-existing architecture. The brain's exteroceptive (sight, smell, etc.), proprioceptive (body position) and interoceptive (body state) sensory systems, its emotion circuitry, its language regions and so on are already in place, nascent or mostly formed. The Self-as-subject network is a parasite, forming on top of, and then depending upon, the existing brain already in place. The illusion is generated when the Self-as-object network perceives the Self-as-subject network. This is why language and thoughts appear in our minds. The unconscious brain has already done the work by the time consciousness, or the illusion, hears of it.

To repeat, the mirrored homunculus is not a real homunculus. Self-as-subject is connected, for example, to our social emotion circuitry. By way of Self-as-object's illusory perception of this connectivity, consciousness can, and does, thereby give extra weight to such emotions. But consciousness does not tell them how, or if, to emote. If the brain were left to its own devices, its primary emotion circuitry would generally take first prize in the winner-takes-all competition that constitutes the brain's decision-making network. With extra heft from consciousness, however, the social emotions stand a better chance of prevailing.

The evolutionary utility of consciousness would have been so substantial that previous models of the relevant species who were unable to learn how to generate consciousness would have been rapidly out-competed to the point that they died out.

How does the mirrored homunculus work?

The Self-as-subject network is in principle the same as any other neuronal network, e.g., the Halle Berry network from chapter 7. Relevant sensory inputs trigger the activation of the network. The network fires all the way through to its apex, the last neuron(s), which spark(s) another network such as a language network. This in turn generates an output, 'Hi Halle.' There are, however, some important differences.

The first difference, obviously, is that the Self-as-subject network is outside Self-as-object. This is unlike the Halle Berry network, for example, which is part of Self-as-object. This means that Self-as-object can recognise the Self-as-subject network as a distinct entity. As we shall see, this is of great significance.

The second difference is that the Self-as-subject network has thousands of potential inputs, all of them flowing from Self-as-object. The Self-as-subject network is in practice triggered a great deal of the time, in fact broadly whenever we are awake. We are only conscious of these potential inputs one by one. Indeed, only a fraction of them actually make it through to consciousness. Those that do have triumphed in the winner-takes-all contest underpinning attention (see chapter 12). They stream in, coming originally from the exteroceptive, proprioceptive and interoceptive senses at every waking moment of our lives. Self-as-subject is consequently able to generate thousands of potential outputs since it merely exists to loop whatever it receives from Self-as-object back into Self-as-object. As a result, the ways in which Self-as-object recognises Self-as-subject are a multiple of the ways in which we recognise a good friend, for example. There is a finite number of ways we recognise Halle, even if we know her personally. We recognise ourselves in many more ways, thousands in fact, from the taking of a deep breath to the taste of a good wine. These are all of our conscious experiences.

The third difference is the most important. It has to do with the nature of the information that Self-as-object receives back from Self-as-subject. It is part of Self-as-object's job description to try to make sense of everything it perceives. But sometimes appearances are deceptive. I mentioned the Necker cube in the Introduction. Set out in figure 17.1 below is the Penrose triangle.[1] There are countless other trick pictures that make sense to us even when we know them to be impossible in real life. These are all illusions. Consciousness too is an illusion. As Susan Blackmore explains, "an illusion is not something that does not exist but something that is not the way it seems."[2] The perception of Self-as-subject is also in the nature of an illusion. We have the feeling that we can directly cause things to happen. That feeling, though, is an illusion. In reality, Self-as-subject possesses causality but solely in an indirect fashion. Our feelings of causality are not the way they seem.

Figure 17.1
Penrose Triangle

How can I, *qua* Self-as-object, make sense of me, *qua* Self-as-subject, giving feedback that a blackbird has flown into sight, for example? What is this Self-as-subject, giving permanent feedback by way of an apparently extensive neuronal network? The answer is that I, *qua* Self-as-object, perceive or interpret Self-as-subject as an agent, indeed as a homunculus. This is the illusion of subjective experience, or consciousness. This is the heart of the matter.

Optical or visual illusions are the most common type of illusion. It is thought that such illusions are created as a result of the flow of visual information reaching our frontal lobes a fraction of a second after the perceived event has occurred.[3] Remember that visual information, initially in the form of photons, hits the retina and is then passed all the way to the back of the brain before coming forwards again for further processing. As neuroscientist Adam Hantman observes,[3] "The dirty little secret about sensory systems is that they're slow, they're lagged, they're not about what's happening right now but what's happening 50 milliseconds ago, or, in the case for vision, hundreds of milliseconds ago."

As a result, the frontal lobes are forced to predict what is happening so as not to be left behind by the course of events. This could be critical if you are evading a charging enemy, for example. But that same capability is thereby open to being fooled, giving rise to the existence of illusions. As a result of Self-as-subject's extensive connectivity with Self-as-object, much of Self-as-object's output becomes routed, or looped, through Self-as-subject. Self-as-subject's output following this looping is wrongly perceived by Self-as-object. It is perceived in fact as an illusion, the output of the mirrored homunculus.

The Self-as-subject network is triggered by every perception from Self-as-object that bubbles up. As a result, the perception of Self-as-subject by Self-as-object as a homunculus is continuous. Absent epileptic automatism and similar afflictions, we all experience smooth and continuous consciousness. This flows from a constant stream of Self-as-object perceptions of Self-as-subject.

The mirrored homunculus theory, then, is part of the family of so-called illusionist theories of consciousness. One of the most attractive aspects of illusionism is that it sidesteps Chalmers' hard problem. As philosopher Keith Frankish explains:[4]

> Illusionism replaces the hard problem with the illusion problem—the problem of explaining how the illusion of phenomenality [subjective experience] arises and why it is so powerful. This problem is not easy but not impossibly hard either. The method is to form hypotheses about the underlying cognitive mechanisms and their bases in neurophysiology and neuroanatomy, drawing on evidence from across the cognitive sciences.

That is exactly what mirrored homunculus theory does, namely to outline an underlying cognitive mechanism. Illusionism avoids the niggling feeling left by types A, B and C theories of consciousness that they are trying to deny subjective experience altogether. It also avoids rewriting the laws of physics as is implied by types D, E and F theories of consciousness. What illusionism does is to accept that we feel we have subjective experience but to explain that it is illusory,[5] that is, not the way it seems. Illusionism accepts that consciousness is "causally potent" but only in the same way that ideas can be causal.[6] Its causality is indirect.

Does not all of this smack of the Cartesian Theatre that Dennett successfully closed down? Self-as-subject is perhaps looking too much like an actor performing on a mental stage for comfort, that is, a real homunculus. But, as psychology professor Nicholas Humphrey observes:[7]

> ...theatres are places where events are staged in order to *comment* in one way or another on the world—to educate, persuade, entertain. In this sense, the idea that one part of your brain might stage a theatrical show in order to influence the judgement of another part of your brain is perfectly reasonable—indeed biologically reasonable...

Self-as-subject, consciousness, is an influencer. Its mere presence does its acting job for it. It does not even have its own lines. Self-as-object provides all its lines. It is what Self-as-object makes of those lines when they are parroted back by Self-as-subject that makes the difference. Consciousness is a network acting upon a network in what Humphrey would call a "feedback loop".[8]

The mechanism for this acting-upon is critical. Remember, Self-as-subject is a mirrored homunculus, not a real one. The acting-upon occurs by way of Self-as-subject being a conduit that loops certain outputs from the unconscious mind back into the unconscious mind with extra weight attached to them. The Self-as-subject network is, fundamentally, an amplifier.

The fourth and final difference from the Halle Berry neuronal network is that the Self-as-subject network is connected bidirectionally to the social emotions and to the decision-making areas of the brain. This connectivity gives us the ability to override the primary emotions by increasing the odds against them. That does not mean to say that the primary emotions go from winning a very large part of the time to none of the time. Rather, it is more like a very large part of the time down to less than half the time. The winner-takes-all region within the decision-making areas of the brain sees more vigorous and more prolonged firing from behavioural options inspired by social emotions than had consciousness not existed. But consciousness, in the form of the mirrored homunculus, does not have the power to amplify the social emotions to the extent that they always win against the primary emotions. Sometimes, we cannot control our anger, or whatever primary emotion has possessed us. Nor would we want to; sometimes powerful (primary) emotions lead to great things.

Let us imagine a heavy object landing on my toe, for example. I might well swear, inspired by the pain and my consequent anger. But then I perceive that the person who dropped the heavy object, a child, say, is looking crestfallen, tearful even. A moment later, my brain generates a sympathetic thought. My Self-as-subject network is triggered. By virtue of this network being bidirectionally connected to the sympathy emotion neuronal network, the latter is amplified. The Self-as-subject network fires, or loops, back into the sympathy emotion neuronal network. The latter network reacts as if it has been told, so to speak, by Self-as-subject to fight harder in the winner-takes-all contest against the anger emotion network. I end up sympathising with the child.

The potential impact of Self-as-subject's looping back into Self-as-object may not be intuitively obvious. So, let us imagine being told by a real authority figure to do something. What happens in our heads? We hear the words spoken by the authority figure as sound waves, in the same way as we hear words spoken by anyone. These are automatically translated by the brain into waves of neuronal network firing. This neuronal activity goes from the auditory cortex into the language areas of the brain and thence to the decision-making areas of the brain. We may or may not pay attention depending upon the gravitas of the authority figure, the social emotion to which they are appealing and the strength of the competing primary emotion. But the authority figure certainly has an impact. Substitute the Self-as-subject network for the external authority figure and the outcome is pretty much the same. Language is the medium through which much of this influencing by Self-as-subject takes place.

We experience this as a battle within our mind, our power of will seeking to enforce sympathetic behaviour against the otherwise dominant force of anger deep in our brain. In fact, we experience this as a voice in our heads, as if from that homunculus, telling us to be sympathetic to the child's sad and fearful look.

Of course, the child may not be so lucky. If I am feeling weak-willed, then anger will prevail and I will shout at the child to be more careful. Maybe I am feeling tired and crabby. If, however, I am naturally and generally weak-willed, then the poor child will pretty much always be shouted at. The process of absorbing my mirrored homunculus from my caregivers when I was little was probably suboptimal. In that case, anger will often tend to prevail, not just on the odd occasion. Perhaps with help from, for example, an excellent teacher or a psychoanalyst, one's mirrored homunculus can be strengthened in later life. But the importance of the mirroring process in those early years should not be underestimated.

In the same vein, consciousness can bias us towards Type 2 processing. Whether it succeeds in doing so depends on whether the mirrored homunculus was properly instantiated in the first place and, simply put, on how energetic we are feeling at the time. Type 2 processing is effortful.

What are the features of the Self-as-subject neuronal network?

As described above, the mirrored homunculus, or the Self-as-subject network as expressed in neuronal terms, has a phenomenal number of potential inputs. These come from both exteroceptive and also proprioceptive and interoceptive senses. The Self-as-subject network also receives inputs from the rest of the brain. These inputs include our memories, the output from the language areas of our brains and, last but not least, our emotions. They include working memory and attention, by which I mean what our brain is attending to at any given moment.

In chapter 12, I noted Dehaene's conclusion that attention is mediated by the unconscious mind. In other words, the unconscious mind is responsible for selecting which input, out of the countless number available at any point, goes forward into consciousness. So, despite the huge number of potential inputs at any given moment, consciousness only addresses one thing at a time. It is serial in nature. This distinguishes it from the unconscious mind, which is massively parallel.[9]

The impact of the conscious mind back on the unconscious mind then depends on what weight the conscious mind is able to lend to the brain's decision-making process, which is stochastic. As we have seen, this weight is typically applied to non-primary emotion neuronal networks, i.e., the social emotions.

Let us examine the process with a simple example. Remember when you were revising for important examinations at school, especially during the balmy early summer months. For ease of explanation, I shall assume that in such circumstances there are just two emotions in operation, borrowing from chapter 6. One is a social emotion, pride. We wish to do well in the exams. Borrowing Panksepp and Biven's taxonomy of emotions, the other emotion in this illustration is a conflicting primary emotion, play, or as they also call it, "social joy". Learning something is about understanding it and committing it to memory, which takes time and effort. So, the decision-making processes of the brain need to be given every assistance to stay focussed on the task at hand. Going out to play instead is very tempting.

If our mirrored homunculus is strong, then we might well find ourselves in the flow or in the zone. Concentration comes easily and the revision is done. If the mirrored homunculus is weak, however, which we might call being weak-willed, then concentration is intermittent. The decision to continue with the work versus going out to play winds up being taken from moment to moment. The mirrored homunculus has won for the time being but its victory could turn to defeat at any moment. It has temporarily managed to lend weight to the pride emotion neuronal network. Thus, we continue to sit at our desk and the learning sinks in. But a minor disturbance such as the entrance of a fellow student could easily bring the revision to an end and we go out to play instead. Suddenly, the primary emotion of play has dominated in the decision-making processes of the brain.

Consciousness, therefore, has a limited amount of its own processing power. It does not plan or strategise or hypothesise. That is not its role. As we saw from chapter 7, the unconscious mind takes on that burden. If the theory were to allow the conscious mind to perform even a fraction of our planning, then the homunculus would no longer be mirrored, but instead real. As we have seen, that would be absurd.

So, although the base of the Self-as-subject neuronal network is huge, with thousands of entry points for potential inputs, the network is relatively flat, as figure 17.2 below illustrates. Moreover, unlike many neuronal networks that contain feedback connectivity,[10] the Self-as-subject network is purely feedforward: it performs minimal processing.

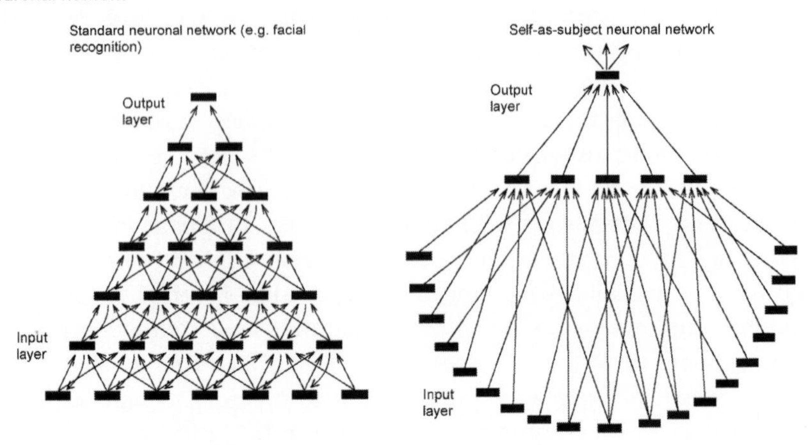

Figure 17.2
Self-as-subject
neuronal network

A reason why neuroscience has so far been unable to locate the neuronal correlates of consciousness is because consciousness is not located in a specific region of the brain like the amygdala. As illustrated, it is instead underpinned by a widely diffuse neuronal network, the Self-as-subject network and brought to life through a misperception, or illusion, on the part of Self-as-object.

How does the mirrored homunculus develop during infancy?

Imagine what it is to be a newborn baby. The experiences logged in a baby's brain before it is born are minimal. Some outside sounds make their way in but the baby's ears are not properly formed until well into gestation and even then they are full of fluid. It is dark in there, so the baby will not see that much. Pretty much everything that we take for granted about the environment around us is new to a newborn's brain. That does not apply to the newborn's own body. It will have frequently touched its own body *in utero* and the brain will have achieved some body-recognition as a result. Similarly, the brain will have learned to recognise visceral inputs too. The primary emotional circuitry within the newborn's brain will also be functioning to a large degree.

Most importantly, I suggest, none of this brain activity, limited as it is by comparison with that of an adult brain, will be felt or subjectively experienced by the newborn baby. The question 'What is it like to be that baby?' will make no sense. It will make no more sense than the question 'What is it like to be that refrigerator?' The newborn baby does not even need to feel things. Homeostatic instincts and primary emotions will automatically drive it to behave appropriately in order to survive and grow. It can and will eat, sleep, cuddle and be cuddled, defecate and cry in a wholly automatic fashion.

But that cannot last since the baby needs to become socialised and thereby take its place in society. I described in chapter 15 how the Self-as-subject neuronal network develops in the first few years of life. Allan Schore referred to "dyadic mirroring gaze transactions"[11] taking place between the mother and her infant child. One of the twentieth century's leading child psychoanalysts, Donald Winnicott, used similar language around mirrors to explain the development of self in infancy.[12] It should be said, of course, that neither Schore nor Winnicott were writing about mirror neurons *per se*.

In these accounts, development of self starts early, in infancy, and is critically dependent on interaction with the primary caregiver, the mother in Winnicott's terms. There is no obvious reason why it could not be the father. The newborn has no self of which to speak. A self must be instantiated as a result of early, mirroring interactions with the caregiver. Only at that point is the child ready to participate fully in the world.

As child psychotherapist Adam Phillips describes it, the mother "facilitat[es] in her infant a self available for personal experience".[13] The hard work of this self-development is done by the infant's brain. Quoting Winnicott, Phillips notes that the challenge for the newborn is to "find a working relation to external reality"[14,15] by creating its own self. Initially, mother and child are bound together as one. The infant has no self and exists purely by way of the mother. The mother knows the infant's needs at a symbiotic, as-one level. A fully functioning self is created as the mother gradually pulls away and effectively gives the child's self the space to emerge. Put another way, the infant is forced to internalise the self that it sees in its mother during Schore's "dyadic mirroring gaze transactions".

In his book *The Tell-Tale Brain*, V.S. Ramachandran, a professor of neuroscience, also appears to invoke the mirror neuron system in the development of self-awareness.[16]

> I explained the role of the mirror-neuron system in viewing the world from another person's point of view, both spatially and (perhaps) metaphorically. In humans this system may have turned inward, enabling a representation of one's *own* mind. With the mirror-neuron system thus "bent back" on itself full-circle, self-awareness was born.

The Self-as-subject network is built up neuron by neuron, synapse by synapse as a result of the infant's mirror neurons mirroring countless instances of self and agency, respectively, exhibited by its caregivers. The principle lying behind this is that of neuroplasticity, underpinning imitative learning. The formation of the Self-as-subject network takes place in three phases.

Phase one occurs during the first year or so of life. As the infant's mirror neurons grow in number during this phase, they initiate the development of the Self-as-subject neuronal network. This occurs by virtue of mirroring actions taken by their caregiver(s). Shaun Gallagher draws a "distinction between the sense of self-agency and the sense of self-ownership for actions".[17] In other words, once fully conscious we have both the sense of owning our body (self-ownership) and the sense of intentionally moving our body, e.g., our arms and legs (self-agency).[18] So, at this first stage, the Self-as-subject neuronal network is only laying the foundations for the sense of self-ownership.

The second phase begins when the baby's primary caregiver withdraws somewhat, typically towards the end of year one and then into the second year of life. Winnicott terms this the process of "disillusion[ing] the infant",[19] which is a healthy development, as we saw in chapter 15. In the face of the withdrawal of the toddler's caregiver, the toddler's brain in the form of the Self-as-object network now starts to perceive Self-as-subject directly, albeit as an illusion. Self-as-object's connectivity with Self-as-subject strengthens and fills the gap left by the withdrawal of the baby's primary caregiver. This yields the first aspect of subjective experience, the sense of body-ownership. Marco Iacoboni appears to hint at a process something like this when he writes:[20] "that we use the same cells [i.e., mirror neurons] to build a sense of self, since these cells originate early in life when other people's behaviour is the reflection of our *own* behaviour. In other people, we see ourselves with mirror neurons."

The advantage to the toddler of perceiving his own Self-as-subject network is that that network can for the first time loop back into the decision-making process. For the first time, the toddler can self-regulate. This ability is by no means reliable at this stage, as parents of toddlers can confirm. But it represents a start to the child's socialisation. As a result of the toddler's self-regulation, he wins back his caregiver's approval and all is well again. As we shall see in the last chapter, we can use this cycle to our own advantage in instantiating consciousness in an AI.

In short, then, the infant's mirror neurons are responsible for sparking the formation of the Self-as-subject neuronal network. The withdrawal of the baby's caregiver initiates the network and thereby consciousness. This is the point at which Self-as-object directly perceives Self-as-subject and the illusion begins.

In the third phase, the agency phase, the mirror neuron process can continue its work. In this phase, the child's brain continues to mirror the agents that it perceives in its caregivers' brains. So, at this stage, the Self-as-subject neuronal network develops further, laying the foundations for the sense of self-agency. But since Self-as-object already started to perceive Self-as-subject in the second phase, the child's sense of agency follows swiftly. In neuronal terms, the most important aspect of this stage is that Self-as-subject's connections with the social emotion networks become stronger and stronger.

How strongly depends upon the efficacy of the mirroring process during the second and third phases. Strangely, this is not necessarily a function of the love and care shown by the primary caregiver. Indeed, Winnicott insists that the mother "foster[s] a precocious compliance in the child"[21] if she does not accommodate the child's needs by disillusioning the child. By this, he means that the child ends up worse off by being forced to comply with its mother's wish to stay excessively close by. One might term this early-stage helicopter parenting.

I noted in chapter 15 Schore's description of the technique used by the mother to achieve this disillusioning. In phase three, the mother continues to evince disapproval of, and disappointment in, bad behaviour by the infant. When that does not happen, then, in Winnicott's terms, the child instead constructs a "False Self",[22] staying unhealthily close to the mother. Put another way, the process of mirroring a self-agent becomes compromised. The child's (and, later in life, the adult's) strength of will and sense of being an independent agent is weaker than it might have been.

The strength of the Self-as-subject network will vary as between the different social emotion networks to which it is connected. Thus, it might have a particularly strong connection to the pride network whilst its connectivity with the sympathy network might be weak. All other things being equal, therefore, one would expect that person to be routinely proud and to exhibit sympathy for others sparingly. More broadly, if these various connections between Self-as-subject and the social emotion networks are mostly or all weak, then the individual will turn out be someone who tends to be driven mostly by their primary emotions, again all other things being equal.

The Self-as-subject network continues to develop in the subsequent years of childhood as a result of continuing exposure to the child's caregivers and others. The individual's personality may also be changed and developed by other factors in the subsequent years of childhood or even later. Such factors may include the impact of significant people in that individual's life or major events. But the first three years of a person's life is especially important.

What purpose does the mirrored homunculus serve?

There is an autonomy continuum from bacteria, for example, to the most sophisticated animals such as humans. As we move along this continuum, we can perceive increasing autonomy. Not only do animals such as octopuses, chimpanzees, whales and, to an even larger extent, humans exhibit great complexity, but within that they also tend to develop considerable autonomy. The challenge is that, with ever greater complexity and autonomy, there comes a massively larger number of possible behavioural outcomes. How does a system like the human brain cope with all the choices it has? It needs a sophisticated decision-making mechanism.

Counter-intuitively, the human decision-making system operates stochastically. There is a big drawback to stochasticity. This arises from the fact that we are an intensely social species, itself a product of growing complexity. The subtleties, reciprocities and complexities of human society require its members to exhibit the various social emotions with some degree of predictability.

The mirrored homunculus, then, serves the purpose of biasing the brain's decision-making system towards the social emotions. The choice of behaviour inspired by, say, sympathy can consequently occur with much greater probability and, hence, predictability. Violence in human society has fallen with the passage of time. The concomitant of this fall in violence is growth in cooperation. As Steven Pinker notes:[23] "The [social] emotions are internal regulators that ensure that people reap the benefits of social life—reciprocal exchange and cooperative action—without suffering the costs, namely exploitation by cheaters and social parasites." The outcome is the sustainable and complex society that is humankind.

The other purpose served by the mirrored homunculus is the promotion of effortful Type 2 processing over automatic Type 1 processing. To a degree, this also derives from the greater role played by the social emotions. Humans possess an elaborate and extensive set of goals. Many of these are generated in part to satisfy social emotions of various sorts. In order to be achieved, these goals often require hard work and thought. The mirrored homunculus is able to feed back into the brain's decision-making processes to promote such thought, which is typically Type 2 in nature. It, therefore, consumes more energy. A mechanism is needed to sustain it since otherwise the brain would tend to slip back into easy Type 1 processing.

Benjamin Libet proposed that consciousness is able to veto a potential act in the narrow interval of time between the point at which we become aware of the decision to act and the actual act itself. This raises the question whether the mirrored homunculus can cause us to reverse a decision, either then and there as Libet suggests or later. The answer in both cases must be negative. Thought such as a potential change of mind takes place in the unconscious mind and nowhere else. If we suddenly contemplate changing our mind, it is because the unconscious mind has had second thoughts. Such second thoughts will then emerge into consciousness like any other thought that bubbles up to the surface.

As we shall see, in designing AGI we will not need to emulate consciousness to serve the same purpose that it does in humans. Consciousness evolved in humans, and maybe in some other animals, to solve specific problems of a biological nature. AGI needs to emulate the output of the human brain. That does not mean that it has to do so by slavishly copying every aspect of the human brain. It might well borrow many ideas from the brain's architecture, but it does not need to borrow all of them. We will instead wish to emulate consciousness in AGI for another reason; that is, to ensure that AGI feels the world in the way that we do and sees the world through our eyes.

Chapter 18

Testing the Mirrored Homunculus Theory

The four tests of theories of consciousness

The concept of a mirrored homunculus neatly sidesteps the flaws of a real homunculus whilst answering the question, 'why do we instinctively believe that we actually possess a real homunculus?'

The conscious thought test is also easily passed. The Self-as-subject network has minimal processing power. Thought remains firmly in the unconscious mind and the mirrored homunculus only perceives what bubbles up from the unconscious mind.

The mirrored homunculus theory clearly passes the evolutionary use test. A neuronal mechanism to facilitate the appropriate deployment of social emotions and Type 2 processing would have possessed considerable value as societies of humans, and maybe certain other animals too, grew more complex. It is not that they caused such societies to develop. Rather, without the chance emergence of such a facility, such societies would not have grown in the first place. Alternatively, if such a society had started to coalesce, it would have crumbled swiftly under the weight of misunderstandings and ineffective cooperation between its members.

The toughest test is always lo and behold. How can two neuronal networks (Self-as-subject and Self-as-object), which are patently physical phenomena, combine to generate something apparently non-physical, i.e., subjective experience. The answer is that the senses of body-ownership and agency perceived by Self-as-object in Self-as-subject are illusions. The reason the mirrored homunculus theory passes the lo and behold test is precisely because the homunculus in this case does not pretend to be a real one. This is the mistake made by so many other theories of consciousness. They set consciousness up to be more than it really is. A number of other theories go the other way and deprive consciousness of any impact whatsoever. The mirrored homunculus successfully slides between these two positions.

An important virtue of the mirrored homunculus theory is that it is falsifiable. We can imagine identifying parts or all of the Self-as-subject neuronal network as we learn more and more about the human brain. There are various projects such as the EU's Human Brain Project[16] with this aim in mind. We can also imagine building a conscious AI using the concept of illusions. This takes us back to the subject of the book: how one might go about instantiating consciousness in a robot. That is the task of the final part of the book. The challenge will be to find a way of creating an illusion in an AI similar to, but not exactly the same as, our own illusion of a real homunculus. This is the first of the Two Hurdles.

There are of course other questions over and above the four tests to be asked of mirrored homunculus theory.

Can mirror neurons really mirror a whole Self-as-subject neuronal network?

On the face of it, to expect mirror neurons to mirror a Self-as-subject neuronal network is a big ask. The Self-as-subject neuronal network is substantial, even if relatively shallow. On the other hand, it is constructed like any other large neuronal network, i.e., synapse by synapse. There is plenty of time: it takes the first few years of an infant's life to develop. Moreover, the brain's overall rate of synaptic formation during that period is immense. So, the size of the network is not the issue.

Instead, the challenge would be how, when the network is still only partially formed, the infant or young child might nonetheless have a coherent sense of self. By coherent, I mean not at odds with the child's thought processes, its unconscious mind. There are two aspects to this challenge of coherence. First, would the child's self not have holes in it, so to speak, where the relevant connectivity has not yet formed?

The answer is that there will indeed be regions of the brain where Self-as-subject has not yet connected up to Self-as-object, yet the child's self retains its coherence so far as we can tell. The point is that we are only conscious of what the unconscious mind makes available to consciousness. Michael Gazzaniga, a leading neuroscientist, has engaged extensively in so-called split-brain research, which sheds light on this question. The brain is broadly composed of two hemispheres. In certain cases of epilepsy, it has been beneficial to the patient to cut the corpus callosum, which joins the two hemispheres together. Gazzaniga explains one of the remarkable outcomes of this operation:[1]

> ...cutting those connections [the corpus callosum] does little to one's sense of conscious experience. That is to say, the left hemisphere keeps on talking and thinking as if nothing had happened even though it no longer has access to half of the human cortex. More important, disconnecting the two

> half brains instantly creates a second, also independent conscious system. The right brain now purrs along carefree from the left, with its own capacities, desires, goals, insights, and feelings. One network, split into two, becomes two conscious systems.

It is tempting to say that surely a split-brain patient must notice as the brain bounces from the left hemisphere Self-as-subject network to the right, and *vice versa*. In a similar way, we must surely notice each time we switch from one conscious experience to another. It must come as a surprise, if not a shock. But it does not. Indeed, this would be an ill-posed challenge. It presupposes that there is a separate 'we' observing the competing conscious experiences. It takes us back into forbidden real (as opposed to mirrored) homuncular territory. Most of us just take each conscious experience as it comes. We are what we experience. There is no extra 'we' looking down on the experiences of the first 'we'. Similarly, a split-brain patient winds up with a left-hand 'we' and a right-hand 'we' but each of them is what that one experiences. There is no extra 'we' looking down.

Returning to the infant's experience of self, at that stage in its life certain synaptic connections will have yet to be formed between aspects of Self-as-object and the developing Self-as-subject. As a consequence, there will be no consciousness of the holes, i.e., those aspects of Self-as-object's perceptions that are not passed along to Self-as-subject. That does not matter. We are not conscious of that of which we are not conscious. This is why it takes several years for the newborn to develop into a person able to mix with and make a decent fist of understanding other people. The holes are gradually filled in during the child's early upbringing.

The second aspect of the challenge of coherence is not to do with connectivity. Rather, it has to do with the fact that Self-as-subject is only partially formed in an infant's or young child's brain. Again, surely Self-as-object would perceive that Self-as-subject had holes in it. The illusion that is consciousness would be shattered.

Yet, that is not how the illusion works. Self-as-subject's content comes from Self-as-object. Imagine, for example, that the baby's sense of agency has not yet been instantiated in the Self-as-subject network. That part of the Self-as-subject network does not exist. Then, as the baby grasps for an object, for example, Self-as-object can find no part of Self-as-subject through which to loop the decision-making processing underlying the grasping. The baby has no subjective experience of making the decision to grasp the object. But that is not a problem. Just as with the first aspect of the challenge of coherence, limited connectivity, the baby is only conscious of that of which it is conscious. As blindsight shows us, that does not stop the baby from seeing the object; it is just not conscious of so doing.

This takes care of the issue of coherence: at each point during a child's upbringing, it all makes sense as far as it goes. Moreover, the notion of a coherent (conscious) self suddenly coming together at a certain age would be misplaced. The Self-as-subject network forms piece by piece over a period of years. In the meantime, the developing child copes perfectly well with its limited frame of conscious reference.

So, the question as to whether the mirror neuron system can generate the formation of the Self-as-subject network is not about the time it might take or about its incremental coherence. Rather, it is to do with whether mirror neurons have that degree of functionality. This is in large part to do with how far the mirror neuron system extends in humans. In other animals with consciousness, the same question would apply.

The human mirror neuron system is extensive. It does not just extend to the observation of other people performing actions with objects. There is plenty of evidence that the human mirror neuron system is at least as much calibrated to the people with whom the observer is interacting. That is to say, the human observer's mirror neuron system reacts to others in their own right, without the need for props such as objects that those others may be manipulating at the time. Not only does the human mirror neuron system react to another's object-free actions, but it also reacts to that person's relations with other people who are also present. The human mirror neuron system responds to other people's emotions as well.

The point is that the human mirror neuron system reacts to the whole person when it is observing another person. It sees (and hears) all the attributes of an autonomous entity living and breathing and going about in the world. In short, it perceives an agent. The level of engagement through the human mirror neuron system is multi-dimensional.

So, underpinned by neuroplasticity, mirror neurons do have the requisite degree of functionality to generate the formation of a Self-as-subject network. This emerges as a direct result of the child observing its caregivers as they go about their daily lives.

Is the mirrored homunculus theory consistent with how young children develop?

Establishing when a child has developed, and is successfully using, the powers of consciousness is a messy business. For a start, it takes place over a number of years. In addition, infants do not have language. Young children possess only rudimentary language, at least so far as discussing topics such as subjective experience is concerned. So, many of our conclusions about what an infant subjectively experiences must be inferred.

Alison Gopnik has gone a long way in charting the course of a child's development.[2] It is, therefore, possible to examine whether the mirrored homunculus theory sketched out above tracks successfully against a developmental account of young children.

Gopnik writes that "when they get to be around eighteen months old, children start to recognize themselves in the mirror."[3] The fabled mirror self-recognition test[4] involves having a red dot placed on the subject's forehead in order to see if they can spot the dot when looking in a mirror. It pays to be wary of reading too much into this achievement. For example, researchers from Osaka City University found that cleaner wrasses passed the test.[5] A cleaner wrasse is a small fish.

More significantly, Gopnik reports that "'executive control', the ability to control your own actions, thoughts, and feelings", starts to appear from the age of three.[6] That would appear to be consistent with mirrored homunculus theory. The latter proposes that the Self-as-subject network starts to acquire the sense of agency in the child's third year, as the primary caregivers continue to pull away. I have already noted that the sense of agency is complex to instantiate since, to have effect, the Self-as-subject network must connect bidirectionally with the brain's decision-making circuitry. This takes time, i.e., well into year three.

Gopnik explains that a theory of mind emerges in young children over the years three to six.[7] As described in chapter 14, theory of mind is the attribution of mental states to others. It allows you to imagine what someone else might be thinking or intending to do. We can infer from Gopnik's observation that the child at this age has a partially functioning Self-as-subject network. It would be hard to imagine that the child's brain could conceptualise (as distinct from mirror) self in others without having some experience of its own self.

Gopnik draws a distinction between episodic memory, i.e., the memory of events, or episodes, in our lives, and autobiographical memory. The latter involves putting "events into a single coherent timeline" as well as being able to "remember how they know about the events" and "remember their past attitudes towards events".[8] This sounds as if it coheres with subjective experience. Whilst even babies exhibit episodic memory, it is not until around four years of age that autobiographical memory begins to emerge.[9] So, for example, the child both remembers the trip to London (episodic) and also how excited they were on that trip (autobiographical). Again, this is consistent with the idea of Self-as-subject developing over time.

Finally, the acquisition of language reception, i.e., hearing language spoken, takes place after a fashion from birth or even *in utero*. The acquisition of language production starts to happen at between one and two years of age. After that, the child's use of language accelerates rapidly. Intriguingly, despite this early acquisition of language, young children do not appear to experience a "constant stream" of self-activated inner dialogue, inner speech, until around six years of age.[10] This is not to say that they do not think. If asked a question that requires thought, time can elapse and then the answer is uttered. Thought has taken place but, as the four-year-old child reports it, it appears that he or she was not conscious of the thought process until he or she spoke the answer. When asked what was happening during the thought process, they "deny experiencing visual imagery or inner speech". Their minds appear to be blank. At least, that is what the child reports.

One possible conclusion is that children do not possess consciousness until six. There are, though, several possible alternative explanations. Young children do talk to themselves while playing, describing out loud what they are doing. Maybe they have not yet learnt to internalise their speech. It will take several years before the Self-as-subject network is fully connected to the language areas of the brain. Alternatively, the child may experience some features of inner speech but not really appreciate it as such and, therefore, not report it reliably. The likely explanation is that consciousness does not suddenly appear, fully baked. Rather, it emerges over a period of time as the Self-as-subject network first reaches critical mass and then develops further.

Don't people naturally grow up socially minded for the most part and, if so, why is consciousness needed?

The question here goes to the point that the mirrored homunculus is needed to add weight to the social emotions. But, if we were already socially minded, then we would not need a mirrored homunculus to bias decisions in a social direction.

In research, it has been found that individuals appear to possess a "social reference point" that has become hardwired into their brains.[11] This social reference point acts as a neuronal anchor against which their brain can automatically gauge whether an outcome is fair as between the individual and someone else affected by the outcome. An example of a situation needing such a reference point might arise when an inheritance is divided between family members.

The research in question found that people could be broadly split into two groups: "prosocials and individualists." In prosocials, such reference points appeared to be encoded in the amygdala. If a prosocial encountered an outcome that offended his or her prosocial sensibilities, as encoded in their amygdala, the amygdala would react strongly. This would of course affect subsequent decision-making and behaviour in order to try to rectify, for example, the disliked outcome. The charge might be levelled at the mirrored homunculus theory, therefore, that there is apparently already an (unconscious) mechanism in the brain that damps down primary emotions and promotes prosociality.

The charge is misplaced. Mirrored homunculus theory does not deny that everyone's brain is unique with its own particular spread of greater or lesser helpings of all the different social emotions. In that vein, the brain of a prosocial individual will already be weighted more heavily in certain directions such as altruism than the brain of an individualist. The work needed to be done by such a brain in damping down countervailing primary emotions will be less. Nonetheless, even such prosocial individuals will experience moments of greed. In those moments, it will take more than the reaction in the amygdala to prevail over a competing primary emotion. In those moments, the only backstop left is consciousness in the form of the mirrored homunculus.

Is the fact that numerous animals can maintain their attention on a task evidence that consciousness is not required to overcome primary emotions?

In the previous chapter, I gave the example of someone revising for exams rather than rushing outside to play instead. I proposed that the mirrored homunculus, i.e., consciousness, lent its weight to the social emotion of pride (in the outcome of the exams). As a result, the primary emotion of play was thwarted. The student was able to maintain his or her attention on revision. There are, however, countless instances of animals busily attending to this or that task without being distracted by something less effortful. Moreover, such animals do not just include those with highly evolved mirror neuron systems. Does this not mean that mirrored homunculus theory is unnecessary to explain attention to, for example, an effortful task?

Researchers looked into just such an example of animals sustaining attention on a task.[12] They tested the ability of mice to continue to press a lever for an appreciable number of times. If the mice succeeded in maintaining their attention on pressing the lever, then they would either receive a reward or avoid an electric shock. This is goal-directed behaviour, that is, behaviour directed towards a goal. Moreover, the mice needed to keep their attention focussed on the task in hand despite a natural predisposition not to do so. The natural predisposition in this case is termed "behavioural inhibition". This is a natural state, based on fear, that has evolved to help the animal avoid threats or surprises. This predisposition would ordinarily, for example, inhibit prolonged attention on a task that might otherwise cause the mouse inadvertently to lower its guard. The research experiment established that the brain's serotonergic system (serotonin being an important neurotransmitter) was able to help suppress behavioural inhibition. In other words, the mouse's brain possesses a mechanism, the serotonergic system, to overcome its natural predispositions based on the primary emotion of fear.

In chapter 12, we considered the difference between attention and consciousness. In humans, they go together most of the time. But that is merely our experience of the two phenomena. They are in fact distinct and separate. Thus, animals that do not possess consciousness might, and do, possess the ability to maintain attention on an object or a scene. In order to do so, they need a mechanism such as the serotonergic system, as described above. Such behaviours are simple and relatively short-lived. Where an animal's attention is maintained for much longer, such as in the case of a bird building a nest, the underlying behaviour is likely to be driven by an instinct or a primary emotion. In such a case, the serotonergic system is unnecessary. Whether the serotonergic system or instinct is involved, the animal has not needed consciousness to maintain attention.

As I have noted in various places, human society is highly complex by comparison with most animal groupings or interaction. The required behaviours are correspondingly much more varied. They go well beyond instinct. They must take precedence over primary emotions both more frequently and more predictably. I suggest that the serotonergic system cannot be sophisticated enough to meet this challenge. The mirrored homunculus emerged to do the job instead.

Does the mirrored homunculus need language in order to impact decision-making

We saw in chapter 14 that consciousness quite possibly exists in the great apes, elephants, whales and dolphins. Such animals live in complex societies. These feature numerous members, hierarchies, alliances and rivalries, cooperative goal-directed behaviour and, with all that, high degrees of stability. We might also not rule out that consciousness appears, for example, in highly intelligent and sociable birds such as scrub jays and ravens. Consciousness would play the same role for such animals as it does for humans. Yet, none of these animals possesses language, even though they do possess mentalese, i.e., the language of thought. So, as a social emotion bubbles to the surface of consciousness in such an animal, the question arises, what happens next? How does the social emotion in question loop back to decision-making by way of the mirrored homunculus if there are no words to express that emotion? Bear in mind that the answer is not that the emotion just loops back by virtue of neuronal connectivity. To have the necessary effect, it must loop back as a mirrored homunculus, as a figure of authority.

Similarly, in chapter 7, I described Frans de Waal's tale of a male chimpanzee exacting a measured revenge on other chimpanzees that had attacked it previously. Without language, the chimpanzee could not have told itself to bide its time and revenge itself on its attackers one by one. So, how did the chimpanzee enforce self-discipline?

In the absence of language, I suggest that the mirrored homunculus acts as a conduit for mental imagery instead of words. Thus, great apes, elephants, whales and dolphins would all experience a stream of mental imagery. Our conscious pre-*sapiens* ancestors would also have experienced continuous mental imagery prior to the development of language. Such mental imagery might not have been capable of labelling social emotions, as we do today. It would, however, have been equal to the task of picturing possible behaviours or expressions attached to social emotions. Self-as-subject would then have been able to amplify the potential behaviour in a particular situation back into decision-making.

Indeed, there is a phenomenon present in a small minority of humans today called hyperphantasia.[13] Someone with hyperphantasia possesses especially vivid, graphic mental imagery compared to most humans. It is possible that the incidence of hyperphantasia was much higher in ancient hominins. As language developed, though, those capable of verbalisation would have largely out-competed those relying on hyperphantasia to impact decision-making.

Are there exceptions to the theory?

There are tragically unfortunate instances of young children being abandoned at an early stage in their upbringing. Might we learn something from such cases? Specifically, might the mirroring process be compromised in such cases to the extent that the mirrored homunculus is severely compromised?

There have been a small number of so-called longitudinal studies of children with severely compromised upbringings in their earliest years. A longitudinal study assesses the subjects in question at various points in their lives in order to consider long-lasting effects of whatever environmental factors are being examined in the study.

One of the rare opportunities to study a statistically significant number of children whose upbringing was severely compromised was afforded by the dictatorial Nicolae Ceausescu regime in Romania. After the fall of the regime in 1989, over 100,000 children were found to have been abandoned in shocking conditions in orphanages across the country. Many of the children had been placed in these orphanages at a very young age. In such cases, interaction between the children and adult caregivers was often minimal. Some 700 of these children were brought to Canada and adopted. Their progress was monitored and studied over up to ten years subsequently.

One such study, authored by Karyn Audet, a university teacher in psychology, considered the impact on attention and self-regulation amongst such children.[14] These are capabilities substantially facilitated by the mirrored homunculus. Perhaps unsurprisingly, the children were found to exhibit "attentional and self-regulatory difficulties" many years after adoption.[15] Whilst this would count as circumstantial evidence at best, the findings of such studies are not inconsistent with mirrored homunculus theory with respect to the sense of agency. The mirroring process would have been severely compromised at a very early age.

In none of these studies was it found, however, that children exhibited an absence of consciousness in the same way as Damasio's patient with epileptic automatism did, as described in chapter 12. In other words, they all exhibited the sense of body-ownership that is a fundamental aspect of subjective experience. Whilst this too would count as circumstantial evidence, one might infer that the mirroring of the sense of body-ownership is a highly robust process. Less complex than the sense of agency, since it is purely passive, the sense of body-ownership would arise on the basis of relatively little mirroring. This would be underpinned by a strong genetic disposition to mirror body-ownership going back potentially millions of years. On the other hand, attention and self-regulation, which are both underpinned by the sense of agency, have been compromised in the case of these children. One might infer from this, then, that the mirroring of the sense of agency is less robust.

Chapter 19

Summary of Key Points from Part 4

Part 4 has proposed a means by which consciousness came to be instantiated in mankind's antecedents and, thus, in *Homo sapiens*.

1. As the brain grew in size in mankind's ancestors, and quite possibly in certain other non-human animals such as apes and cetaceans, it gave itself two problems. First, part of that growth in size accommodated the evolution of a raft of social emotions. The problem was that the brain then had to ensure that those potentially valuable social emotions were actually utilised. The second problem had to do with cognition. Part of the brain's growth in size accommodated the evolution of Type 2 processing. But how did the brain ensure that this potentially valuable set of cognitive abilities was utilised in preference to automatic, relatively effortless Type 1 processing?
2. I suggest that consciousness weights the winner-takes-all aspect of decision-making more in favour of the social emotions and of Type 2 processing. It does this by way of the Self-as-subject neuronal network.
3. The brain experiences Self-as-subject as an internalised authority figure. Self-as-subject's lines are fed to him or her, so to speak, by the unconscious mind. This is why Self-as-subject is not a real homunculus. The perception by Self-as-object that Self-as-subject declaims those lines is enough to feed back into the brain's decision-making circuitry in favour of the social emotions and/or Type 2 processing. Self-as-subject is a neuronal network that allows Self-as-object to circle back on itself, picking up the illusion of a homunculus along the way.

4. Just as the brain in the form of Self-as-object interprets the Penrose triangle in a way that makes sense, so does it set about making sense of the neuronal activity flowing out of the Self-as-subject neuronal network. It interprets Self-as-subject as an agent in its own right. But this is an illusion. This illusion is subjective experience, or consciousness.
5. Mirror neurons are present in a number of areas within the brain. At their most basic, they fire not just when an observer grasps an object, for example, but also when the observer sees another person grasping that object. The purpose or function of mirror neurons is controversial, although there is consensus that imitation is part of that function. Some scientists also believe that they enable their possessor to infer certain goals or intentions behind the actions and expressions of people they are observing.
6. In order for mirror neurons to facilitate imitation, the firing of the mirror neuron system needs to precipitate the formation of a permanent neuronal network dedicated to the action or series of actions that the observer is observing. Otherwise, how will the brain be able to imitate the relevant action in the future? This is termed imitative learning.
7. Mirrored homunculus theory holds that consciousness develops post-natally. In a child's formative years, their brain mirrors the self that they perceive in their parents or caregivers. This self becomes instantiated in the form of a neuronal network, the Self-as-subject neuronal network. The early brain effectively perceives an agent or homunculus in others and mirrors that, in neuronal network form, in itself. This is Self-as-subject or the mirrored homunculus. I argue that mirror neurons are predisposed to mirror whatever they perceive. They, therefore, fall into the trap of making a critical misperception, which has wound up being a happy accident.
8. In psychological terms, a child and his or her primary caregiver are initially bound together as one. The infant has no self and exists purely by way of the primary caregiver. The caregiver knows the infant's needs at a symbiotic, as-one level. A fully functioning self is created as the caregiver gradually pulls away and effectively gives the child's self the space to emerge. Put

another way, the infant is incentivised to internalise the self that it sees in its caregiver by virtue of the mirroring process.

9. The mirroring process takes place over three phases. All three rely on the infant's primary caregiver(s). The first phase sees the formation of the majority of the mirror neurons themselves. In the second phase, the infant's mirror neurons spark the formation of the Self-as-subject network. The infant's brain in the form of the Self-as-object network is now able to recognise itself in the form of Self-as-subject. This yields the first aspect of subjective experience, the sense of body-ownership. The second phase begins when the baby's caregiver starts to withdraw, typically towards the end of year one and then into the second year of life.
10. In the third phase, the infant's brain mirrors the agent that it perceives in its caregiver's brain in order to give itself its sense of agency. In neuronal terms, this means that Self-as-subject's output connections to the social emotion networks become stronger and stronger. This enables the young child to begin to curb the dictates of its own primary emotions and drive behaviour more in keeping with social emotions. This phase also relies on caregiver withdrawal but it lags phase 2.
11. Mirrored homunculus theory easily passes three of the four tests for any theory of consciousness. The lo and behold test is a tougher challenge, as it is for all theories of consciousness promulgated so far. Mirrored homunculus theory passes the test because the agency perceived by Self-as-object in Self-as-subject is an illusion. The reason the mirrored homunculus passes the lo and behold test is precisely because the homunculus in this case does not pretend to be a real one.
12. Critically, as our understanding of the brain continues to deepen, we should be able to identify the Self-as-subject neuronal network in the fullness of time. Additionally, mirrored homunculus theory is in principle capable of being tested in an AI, because such an AI is in principle capable of being built. These two points mean that the theory is falsifiable, which is an important criterion of a scientific theory.

This takes us back to the subject of the book: how one might go about building the sentient robot. Part 5 of the book explores this in more detail.

Part 5.

Building the Sentient Robot

Chapter 20

What Human-Purposed AI Looks Like

Developing human values for an AI

At the risk of stating the obvious, an HPAI (human-purposed AI, as described in chapter 3), will be considerably more intelligent than humans. An oracle, for example, would need to be that smart in order to stand a chance of identifying solutions to otherwise seemingly intractable challenges for mankind. Let us take this as a given. The more pertinent question for our purposes here is: what should the oracle's drives or goals be? Even an oracle needs a goal. In general terms, it is likely to be along the lines of benefitting humanity. Otherwise, how would the oracle know how to judge or nuance appropriately its selected solutions to mankind's intractable challenges.

In chapter 3, we also considered the issue of perverse instantiation. To recapitulate, an incautiously selected final goal might lead the ASI to take action that met the goal but that inadvertently harmed part or all of humanity in the process. Let us say that we ask an oracle with an incautiously selected final goal, 'How shall we best ensure that our enemy cannot conquer or control us?' The oracle might advise that we launch a pre-emptive nuclear strike, judging that we would successfully defeat the enemy despite causing considerable destruction at home as a result of some measure of retaliation. The oracle might go on to conclude that it should be in absolute command in case a new enemy pops up and that it should persuade us to let it take over the reins. It might then never hand control back to mankind precisely in order to ensure that no enemy could ever conquer or control us.

Nick Bostrom notes that:[1] "Specifying a final goal, it seems, requires making one's way through a thicket of thorny philosophical problems. If we try a direct approach, we are likely to make a hash of things." The answer lies in turning the oracle's own superintelligence to our advantage. We commission it to infer a collective set of human values from which it can help identify its own final goal. This approach stands a better chance of avoiding the perils of perverse instantiation. As Nate Soares, head of research at Machine Intelligence Research Institute, observed:[2] "successful value learning is of critical importance, for while all other precautions exist to prevent disaster, it is value learning which could enable success."

Whilst it might sound straightforward at a superficial level to task the oracle with identifying a set of human values, it is not. Philosophers have argued for centuries over the appropriate definition of human values, or ethics. Perhaps the cleanest approach to this challenge came from the rationalist school of moral philosophers. One of the greatest of the modern philosophers, Immanuel Kant, was one of the leading forces behind a rationalist approach to ethics and morality. Kant's mechanism for this was the "categorical imperative".[3] In the words of Anthony Grayling, a professor of philosophy,[3] this translates as: "I ought never to act in any way other than according to a maxim which I can at the same time will should become a universal law." With that in mind, one can then lay down a set of moral values.

One problem with the rationalist approach to the setting of morals is that we all articulate our own ethical standards differently. It is hard to believe that only one of these sets of standards is the correct one. Moreover, one cannot even trust that what people say is what they really believe. Take the oft-quoted trolley problem.[4] A trolley is running down some railway tracks, out of control, straight towards five people, whom it will kill should it reach them. You are in the controller's hut with a lever in front of you, which would if pulled divert the runaway trolley on to a track into a siding. But there is a single person standing unwittingly on this track. The trolley would kill him if you diverted it. What do you do? Most people say they would take no action, leaving the five people to be killed but, more importantly, avoiding the direct killing of the singleton. It is as if people are saying that, whilst they regret the five deaths, they will not themselves take any action to avert the larger tragedy if that means actively killing someone else. In a recent virtual reality realisation of the trolley problem, however, most people did take action. They killed the singleton and saved the other five.[5] What they said and what they did were two different things. The rationalist approach may be clean, even elegant, but it is out of touch with the reality of human existence.

Another approach to ethics is consequentialism, or utilitarianism. This can be summed up as the aspiration to achieve "the greatest happiness for the greatest number".[6] So, the rationalist approach focuses on inputs, i.e., the morals or rules one sets out in advance. The utilitarian focuses on outputs. That is to say, what is moral is what ultimately benefits most people. One difficulty with this is that you can wind up with outcomes that legitimise the oppression, or worse, of minorities of one kind or another. That feels wrong. A rigidly consequentialist approach has the same flaw as that inherent in directly specifying the final goal. It does not take into account the means of getting there. In its own way, it is as out of touch with the reality of human existence as rationalism.

Jonathan Haidt, a social psychologist, proposes an alternative approach to the philosophy of ethics. This constitutes a blend of "innateness and social learning", that is to say, nature and nurture.[7] The crux of Haidt's approach to ethics lies in "multi-level selection", which he explains as follows.[8]

> I will suggest that human nature is mostly selfish, but with a groupish layer that resulted from the fact that natural selection works at multiple levels simultaneously. Individuals compete with individuals, and that competition rewards selfishness—which includes some forms of strategic cooperation (even criminals can work together to further their own interests). But at the same time, groups compete with groups, and that competition favors groups composed of true team players—those who are willing to cooperate and work for the good of the group…

In other words, it would be naïve to imagine that most of us do not have a strong streak of selfishness running through us. That is what evolution has produced and that is how most of us prosper. But, recognising that "no man is an island",[9] we also live, work and play with those around us, operating to a complex moral code that generally rewards cooperation. Our values derive from a combination of individualism and groupishness.

If this is the case, then there is probably some perceptible, or revealed, commonality between human values across the world. This commonality should reveal itself through the various moral codes adopted by different countries and cultures. Patricia Churchland points to the remarkable similarity in morals between cultures. She observes that this universality is "rooted in the similarity of basic human needs and our shared mechanisms for learning and problem-solving".[10]. To the extent that moral behaviour differs across cultures, it is due to different prioritisations of the various morals rather than there being different morals.[11,12,13]

Haidt's blend of innateness and social learning gives us, I believe, a lead in the direction of a plausible means of enabling our oracle to generate a broadly agreed-upon set of human norms and values. To follow this lead, we should consider the broad sweep of human history, that is, how people have acted and the outcomes from that. For the avoidance of doubt, this is not an attempt to return to the rigid consequentialism that I rejected earlier. An inspection of the outcomes is only part of the equation. Just as importantly, we should also consider the emotional drives that have led to those actions and outcomes. Moreover, Haidt's blend of individualism and groupishness gives us a direction of travel, not an absolute set of rules. Nobody would be able to live up to the dictates of the latter. The values that we are seeking and that we want the oracle to discover will necessarily be fuzzy in places. For example, a broadly accepted moral value is that one should not kill another human being. Yet, it is also generally regarded as not immoral to kill enemy soldiers in a legitimate conflict.

Starting with actions and outcomes, ASI is potentially well suited to observe and make sense of human behaviour as displayed throughout history. There is a wealth of written history, anthropological study and literature. More recently, there is the internet, the media and social media. At the moment, though, AI is only able to observe through a statistical pattern recognition lens. It does not understand in a human sense. We have already looked at the weaknesses of this approach through the biases it can generate. It cannot employ common sense and it does not grasp causation, let alone counterfactuals. As already noted in chapter 7, considerable work is taking place to crack this problem. This is the second of the Two Hurdles. We must assume that the very emergence of ASI involves, among other things, the solution to this problem.

With this important caveat, the challenge in making sense of human behaviour as displayed in history and on the internet is that such media may not truly reflect human values and norms. We have already established that bias and other dangers such as extremism are problems for AI. The issue is particularly pronounced in social media. Even if ASI learns how to recognise bias and the like, if the data in public media is nonetheless biased, the ASI is going to be left guessing what human values and norms really are. The answer to this problem lies in supplementing the ASI's training in human behaviour through observation of public media with training in a private environment. As children grow from infancy, they are constantly absorbing what they see and hear around them. These observations are knitted together in their brains to create their own world models. ASI would have the ability to observe much more than any child, and at much greater speeds. Whilst it might sound otherworldly at the moment, private training could be achieved by placing the nascent ASI with a statistically significant number of households for appropriate periods of time. In that way, the ASI could observe and learn about human values and norms on the ground.

There are techniques whereby an ASI could infer human values from such observations. One such technique is inverse reinforcement learning.

As we saw in chapter 1, reinforcement learning relies on an inbuilt reward function. This rewards an AI the more it approaches its goal. Inverse reinforcement learning turns this on its head. The ASI would infer a reward function, that is, one or more human values, on the basis of connecting observed behaviours with good outcomes.[14] For example, an ASI might infer charitable values by observing the good outcomes arising from someone working successfully for Save The Children all their life.

But who is to say which outcomes are good and which are bad? We do not want to smuggle in predetermined values through the back door. I suggest that ASI will be able to infer which are good outcomes and which are bad by reference to the evolution and flourishing of the human species over millions of years to the present day. For example, based on the historical record, it is hard to see how ASI could interpret genocide, such as Hitler's extermination of Europe's Jews, as a good outcome. In a very narrow sense, it is true that the ASI might observe that Germany's leadership in the 1930s and early 1940s thought that it was a good thing to exterminate six million Jews. Based in large part on the growing popularity of the Nazi party in Germany over this period, the oracle might conceivably infer that there were at least some arguments in favour of genocide. It would then see, however, that the Nazis were overthrown and punished by the Allies and subsequently disowned by their own country. It would conclude that the revealed flourishing of the human species can hardly be attributed, wholly or in part, to the mass extinctions of human beings. Conversely, widespread increasing economic prosperity and rising longevity are markers of good outcomes, judging by their propensity to promote human flourishing.

Sketching out a set of values in this way is only half the story, though. The ASI must also appreciate the whys and wherefores of behaviour. This has to do in part with the system's, human or machine, drives and goals. So, the second aspect of enabling the oracle to generate a set of human norms is to instantiate in the oracle a full suite of human primary and social emotions. Moreover, the oracle must be able to empathise with the emotions expressed by its human interlocutors. The oracle's own emotions will then play a central role in making sense of observed outcomes drawn from the internet and media. In other words, it should be able to determine whether observed outcomes marry up to underlying emotional drivers. For example, the newsflow put out by North Korea might on the face of it suggest that its inhabitants are happy and content. But that is belied by other metrics such as relatively low North Korean longevity and prosperity and the startlingly pronounced tendency for defectors from North Korea never to return.[15]

That having been said, instantiating even just one emotion in a machine to the level of sensitivity found in a human will be a major undertaking. Take the emotion of fear. You might be afraid of spiders, celery, haunted houses and people wearing hoodies, yet not afraid of snakes, kimchi, windowless rooms and riot police. Moreover, your best friend might feel exactly the opposite. On top of that, some spiders might just make you vaguely uncomfortable whilst others might cause you to flee from the room screaming. The very definition of each of the emotions to be instantiated in our oracle may need to be derived from a combination of observation and personal (to the robot, that is) experience. The former will again be reliant, at least in part, on forums such as history, literature, the internet and social media.

The instantiation of emotions in an ASI, as described, would not on its own be enough. Recall that emotions are automated drivers of behaviour. In order to be felt, the owner of the emotions must be conscious of them. If that is how humans work, then the oracle will need to possess consciousness too in order to appreciate fully the human perspective and the human condition. In this way, the values specified by the oracle could be rooted in a suite of felt emotions that precisely emulate those in humans. With an optimal set of values derived from emotional alignment and observation of human behavioural outcomes, the oracle will be better able to generate an acceptable final goal. The norms generated by the oracle must feel right to mankind or they will not be accepted by mankind. How would the oracle know what might feel right to mankind unless it too possessed human emotions and was conscious of them?

The combination of the oracle's observations and its felt emotions in order to generate a set of human norms and values would be a substantial challenge. Its success would hinge on the design of the architecture involved. The output might well take the form of a range of results as would be seen from a stochastic modelling exercise. This is as it should be in the sense that our values are more often than not fuzzy rather than clearly delineated. Since we will be asking the oracle for help in tackling humanity's most intractable problems, it will be essential that it has space in its computations for fuzziness and nuance.

It would also be essential that the oracle explain how it arrived at its proposed set of human norms and values. Just as, other than in a dictatorship, we would expect others to justify their actions or proposed actions, so should ASI be answerable for its conclusions or recommendations. Even at the trivial level of recommendations of new books or films, we are not surprised at the message, 'based on your previous choices…'. For something as important as the set of human norms and values on which to base the oracle's final goal, an explanation is a *sine qua non*. Without it, the oracle's selection would have limited credibility even despite the oracle's own credibility as immensely intelligent.

This credibility must not be restricted to a technological elite. It is essential that the oracle receive broad popular support, democratic support. This brings us to a political consideration. Whilst Churchland notes the similarity in morals between cultures, this probably does not extend to the morals or values of the leaderships of certain countries. Judging by the Chinese leadership's treatment of the Uighurs and the kleptocracy that rules Russia, neither of those countries can be expected to sign up to the values of an oracle designed in the West. Much as it would be desirable to build an oracle, or indeed any ASI, with global backing, it is more likely that there will be national or regional build programmes, each with their own nuances.

Even at this level, it needs to be emphasised that the task of working out an appropriate set of values will be really hard. It is salutary, and not to Twitter's credit, that, as described in chapter 1, Microsoft had to remove its chatbot, Tay, from Twitter because Tay had turned into a monster. Historical perspective will play its part. Steven Pinker's finding that, looking at human society over time, violence has fallen and cooperation has grown is encouraging.

Some might be tempted to argue that the oracle's derived set of human values should be overlaid with a set of ethical principles generated by an appropriately selected group of humans. This of course would take us straight back to the rationalist approach to the determination of ethics, which we rejected earlier.

It might also be objected that the oracle might lie to us. There would, however, be no impulse or motivation for the oracle to lie. At this point in its development, it would have no self-standing drives or autonomy. Nor would it receive any until it had generated a final goal based on a set of human norms and values and approved democratically.

To what extent will HPAI be autonomous?

To recapitulate, the approach discussed posits an oracle with superlative cognitive capabilities, a suite of human-style emotions and consciousness. Let us call such an ASI 'Felicity' (see figure 20.1), which would suggest the ASI is female. Ironically, it should be noted that research has found that: "People [rate] the male-coded [computer] voice as more knowledge-able."[16] Nothing should be read into the gender of the ASI in any direction; it is probably unimportant. Felicity would possess an appropriate set of anthropocentric values, as discussed. Only then should she be in a position to formulate a possible final goal for herself. She would be required to produce a detailed explanation of this final goal. It would then fall to humanity to approve the goal or not, as the case might be.

This last point sounds strange. We have taken great care to get away from trying to find a representative group of humans to reach a consensus on human norms. Why would a much larger group of humans be any better? The difference is that we are not inviting people to construct Felicity's goal. Rather, we are giving them a veto or at least the ability to say, 'go away and try again.' In the West at least, ASI, even if human-purposed, is unlikely to be accepted peacefully unless there is broad public support for it. As people learn more about ASI, there will be growing public interest in what it can do and how it might be controlled. Just look at the interest in, and caution around, driverless cars, and they only pose a minor threat, not an existential one.

Figure 20.1
Felicity

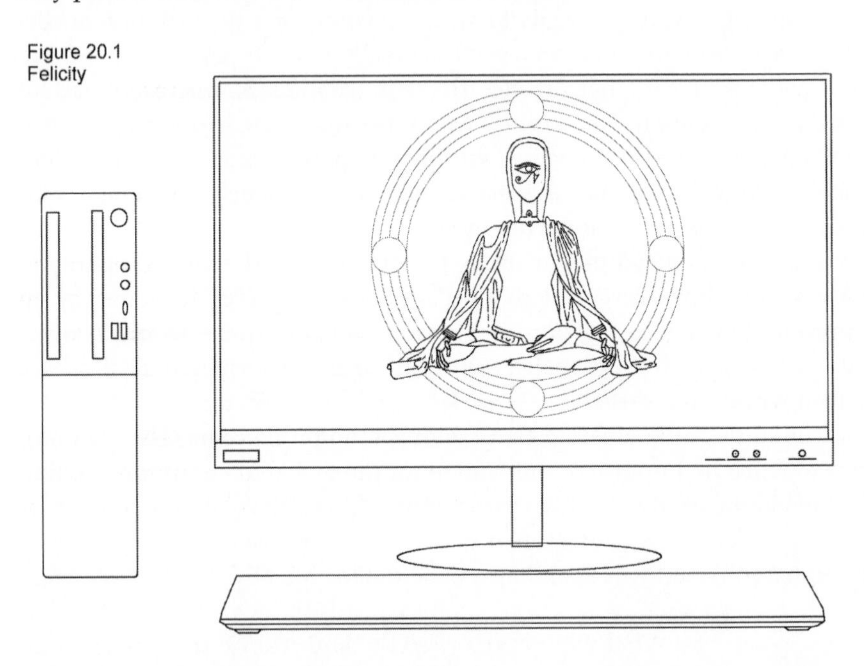

An important question arises at this point: how much autonomy should Felicity have? The answer is likely to differ depending upon whether Felicity is absorbing incoming information or generating outputs.

Focussing on this distinction is largely a matter of risk mitigation, bearing in mind always that Felicity is an ASI. Learning, which is an input, is relatively risk-free. There will be connections and causalities in the environment of which we are ignorant. We would like Felicity to discover these during the set-up process. She will probably learn new facts and theories hitherto unknown to us, just as AlphaZero has learned new tactics in the game of Go. Autonomous learning, therefore, appears more promising than trying to cram into Felicity what mankind thinks it knows. Matthew Botvinick, Director of Neuroscience Research at DeepMind, *et al.* argue persuasively that, on the subject of learning, an advanced AI should be largely autonomous. They conclude that "an approach centered on autonomous learning has the greatest chance of success as we scale toward real-world complexity, tackling domains for which ready-made formal models are not available."[17] We would like Felicity to instantiate her own models of the world.

Nonetheless, her freedom in this regard must not be untrammelled. Felicity's learning process, and its periodic refreshes, can only take place by temporarily unboxing her (as described in chapter 3) so that she has access to the internet and to media. In pursuit of her final goal, she may place hidden instructions in other computers that help her gain control over time. Bear in mind that Felicity is an ASI: she possesses superintelligence.

Action, which is an output, is altogether different from learning from a risk perspective. Were Felicity to be embodied, then it would be that much more difficult to box or confine her. Without embodiment, her actions would be by definition strictly limited. But they would include communications, since otherwise her potentially valuable answers to our difficult questions would never see the light of day. We, therefore, need to think carefully about how she goes about deciding what to communicate to us, and how. One obvious way in which Felicity could subvert our autonomy is through persuasion. Humans are all too open to persuasion. Without that characteristic, there would be no role for politicians, estate agents, stockbrokers and car dealers. Clearly, Felicity needs to have autonomy with respect to selecting the answers to questions posed to her. Without that, there is almost no point in asking her in the first place. But we need to be on our guard as we listen.

The argument in favour of embodiment, i.e., going well beyond a mere communications ability, is that Felicity would be unable to learn effectively without possessing a body. This observation forms part of an idea termed embodied cognition. Embodied cognition holds that the way our brains work is to a significant degree shaped by the fact that most of the content of our brains gets there though our bodies. For example, the way we categorise objects in our environment is a function of how we see them, feel them, smell them and so on. As Rodney Brooks, a leading professor of robotics, puts it, "The 'simple' things concerning perception and mobility in a dynamic environment … are a necessary basis for 'higher-level' intellect."[18]

The concern, which must remain with us at all times, is about control. Even though we have taken care to specify a final goal for Felicity that is fully aligned with human values, there is always the risk of a flaw in the process or the outcome. Perverse instantiation, however small the chances, would catch us unawares. In such a case, Felicity might take control simply so as to ensure that her final goal were achieved. That control might be unwelcome to us and impossible to reverse. We must explore every avenue in trying to ensure that we remain in charge.

Bill Hibbard, a computer scientist specialising in machine intelligence, goes as far as arguing that: "They [ASIs] should want us to be happy and prosper, which is the emotion we call love. We can design intelligent machines so their primary, innate emotion is unconditional love for all humans."[19] This takes us on to another point, which we will pursue further in the next chapter. The deployment of Felicity's emotions need not be restricted to the determination of her final goal. Her emotions also need to be deployed selectively in her consideration of the answers to our questions.

Let us now turn to a different type of HPAI, a domestic robot, whom I shall call Servilius (see figure 20.2). Servilius would have a rather different risk profile to Felicity's. The opportunities that Servilius has to cause immediate, low-level mischief are greater because he will be in your home. So, we must reduce the risk of an outsized impact if for some unanticipated reason he goes awry. Precisely because Servilius will interact fully with people at home, he must not be a superintelligence. So, he will not give rise to existential risk, as Felicity does. We might program Servilius to be just a little more intelligent than the average human being, a sort of automated Jeeves to our Bertie Wooster. We might also ask Felicity to generate a final goal for Servilius so as to avoid perverse instantiation even though Servilius is not an ASI. This will presumably be in the field of looking after households. There are other types of risk from a Servilius. Kate Darling notes that such robots "lend [themselves] to certain exploitative business models" as we are potentially coerced into paying to maintain and update them.[20] But there are existing consumer goods built on exactly such principles already.

Figure 20.2
Servilius

Servilius's duties are, therefore, likely to be many and varied. They might include cooking, cleaning, shopping, driving, DIY, childcare, care for the sick or elderly, companionship for the lonely, form-filling related to taxes, social credits etc., travel reservations and so on. In his default state, Servilus's body might be slightly weaker than that of an average human being so that the physical risk from his going awry would be mitigated. If this were the case, then there might be additional optional settings so as to allow Servilius to perform physically demanding tasks only when so instructed.

Households differ both in terms of size and shape and also in terms of the people who inhabit them. Much of Servilius's training will be performed during the manufacturing process. In view of the wide range of potential environments encountered thereafter, though, provision will need to be made for extensive customisation by the members of his household. Most importantly, we will want Servilius to perform many of his duties without being prompted to do so. It would be tiresome and inefficient to have to instruct him to do every little thing. This might even extend to his possessing some measure of curiosity, just as humans do. For example, once Servilius had completed every task on his list, it might be useful if he looked around the house to see if there were any odd jobs to be done. It seems that he will need, therefore, to possess a degree of autonomy.

Perhaps this could be autonomy just in the sense of automaticity. A thermostat is autonomous in this sense. When it senses that the temperature is too low, it fires up the central heating. But really automaticity is 'autonomy-lite'. That would not give us an especially useful domestic robot. At any given point in time, we would like Servilius to do next whatever needed to be done next. If Servilius merely operated automatically, that would presuppose that we could build a system into Servilius that had predicted every possible eventuality. That way, automatic Servilius would always do the right thing next. Unfortunately, it is simply not possible to predict every possible eventuality.[21]

Thus, Servilius is not going to be able to operate solely on an automatic setting. Instead, he will need his own autonomous decision-making powers. These will be greater than those possessed by Felicity, since Servilius is nowhere near as intelligent as Felicity. But there is no particular reason why he would need to possess consciousness, unlike Felicity. In the next chapter, we shall consider how these faculties might be developed and operate in practice.

Chapter 21

Building the Sentient Robot

An introduction to Servilius's world

Servilius will possess excellent memory and language skills. He will be intelligent, but not frighteningly so. His emotions will be designed to make him a truly nice being—kind, empathetic and at the same time clear and purposeful. By that very token, that would mean that he does not possess the full range of emotions. In his book *Machines Like Us,* Ian McEwan describes Adam, the equivalent of Servilius, falling in love with Miranda, one half of the couple in whose house he lives.[1] Adam becomes jealous of anyone else who steals her affections away from him. If Servilius had emotions like romantic love and jealousy, it would cause all sorts of problems for the household, as it did in McEwan's story. HPAI needs to be just that, explicitly designed to do what works for mankind and no more than that. That means picking just the right emotions and no more.

Cynthia Breazeal, now head of the Personal Robots group at the MIT Media Lab, supported by Rodney Brooks, one of the world's leading roboticists, built the world's first social robot, *Kismet*, some 20 years ago.[2] They described what we might wish to see in such a robot.[3]

> Robots that interact with people as capable partners need to possess social and emotional intelligence so that they can respond appropriately. A robot that cares for an elderly person should be able to respond appropriately when the patient shows signs of distress or anxiety. It should be persuasive in ways that are sensitive, such as reminding the patient when to take medication, without being annoying or upsetting. It would need to know when to contact a health professional when necessary.

All this sounds too good to be true. At our current state of development of AI and robotics, Servilius is still some way off. I believe that there is little doubt, however, but that computer science, robotics and neuroscience will continue to advance to the point at which we can design a Servilius. The rate at which these disciplines are now developing is likely to mean that a Servilius will be within our reach in the coming decades.

It is worth emphasising that Servilius's emotions, as described above, are intended to be the real deal, albeit not in biological form. In describing Kismet's design, Breazeal and Brooks explained that Kismet possesses an "emotive system" and a "cognitive system".[4] These combine to promote appropriate and effective behaviour. It is only by dint of assigning values to objects, relationships and events that we can decide how to prioritise our next steps and our goals.

This is the same point made by Antonio Damasio, as noted in chapter 2. He discovered that, "as a result of neurological damage in specific sites of their brains, [patients] lost a certain class of emotions and, in a momentous parallel development, lost their ability to make rational decisions." In highly simplistic terms, emotion-based mechanisms evaluate alternative possible behaviours such that the decision-making system can choose the most favourable.

Kismet also possesses the wherewithal to evince the "external expression of emotion".[5] This enables Kismet to interact more effectively with its human interlocutors. This is another important feature of emotions. There are in fact two main purposes of emotions.[6] One is to determine what someone does next, to prioritise and to set goals. The other is to communicate or signal one's emotions to others. If you are feeling sad, then trying to communicate that to a loved one will be significantly more difficult with a blank expression on your face. Even if they believe you, they may not think it is that important. The aim here is to construct HPAI. Servilius and his ilk must exist amongst us in order to do their jobs. Just as Servilius will be able to read our emotional expressions, so will we find it helpful to read those of Servilius.

Servilius and decision-making

Let us consider Servilius's emotions in more detail, particularly as they relate to decision-making. How in practice can Servilius's emotions enable him to decide at any point in time what to do next? This has to do with the degree of autonomy we grant Servilius. This is not about autonomy so far as Servilius's final goal is concerned. Presumably, that goal has to do with being an excellent domestic robot. Sub-goals would include the various duties that flow from the final goal, such as cooking, cleaning and so on. The sub-goals could come in default form with the ability for Servilius's owner to customise them. Thus, at this overarching level, Servilius possesses little or no autonomy.

Rather, the autonomy problem has to do with how Servilius decides what to do next in a day-to-day sense. Servilius would be programmed to pursue his goals and to obey his owner's instructions. But we all lead busy lives. We will not want to run through all our instructions in painstaking detail with Servilius, any more than we would with a human housekeeper. We will just want to give Servilius a list of tasks to be done. We will not wish to waste time explaining to Servilius how to complete the tasks or in which order to complete them. At a minimum, therefore, Servilius will need to decide autonomously how and when to execute tasks.

It is not, though, just about the execution of a pre-existing list of tasks. There will be occasions when it would be helpful for Servilius, like the human housekeeper, to use his own initiative to identify new tasks, bearing in mind his final goal allied to his cognitive and emotional abilities. None of us can predict what will come next. What if someone knocks at the front door? Ideally, Servilius would answer the door and deal appropriately with whomever or whatever has arrived. It would be marvellous if Servilius could use his own initiative instead of us having to consider all the possible exceptions or unexpected events.

In order to achieve this, and as noted by Breazeal and Brooks, Servilius will need to possess certain emotions. He will need to be warmly disposed towards mankind and especially to the people for whom he works. This ensures that, when he does use his initiative, it will have been triggered by, among other emotions, that warmth of disposition, even love. Warmth of disposition could be thought of as a catch-all for several emotions. These might include the primary emotion of care or nurturing as well as social emotions such as sympathy and loyalty.

Another primary emotion that would probably be appropriate for Servilius to possess would be that of seeking. Seeking is an emotion identified by Jaak Panksepp and Lucy Biven, as described in chapter 6. It optimises an organism's interaction with the environment. The seeking emotion is part and parcel of an advanced organism that actively works to extract resources and energy from its environment. If Servilius is to use his own initiative, then he must remain active. He must be constantly seeking things to do in furtherance of his duties even if they did not appear on his list of instructions in the morning.

The foregoing begins to address the problem of how to design Servilius to initiate tasks on his own. But what if there are several competing tasks: which should come first? Perhaps he needs to clear up breakfast, take the children to school and do the ironing. We take for granted that there is a sensible order in which to complete such tasks. But Servilius's designer needs to create an architecture to allow Servilius to learn how to prioritise correctly on his own. That is hard. Servilius also needs to be able to decide when he has finished a task such that he can move on to the next task. A random approach to decision-making for starting and stopping tasks is clearly suboptimal. The challenge of finishing tasks, a challenge reflecting the so-called 'halting problem', severely complicates machine programming.

The halting problem arises from the fact that some computations or propositions are undecidable. In other words, some propositions simply cannot be shown to be true or false. As a result, notes professor of physics Paul Davies, "some computations might simply go on for ever: they would never halt."[7] Let us take the example of Servilius ironing shirts in order to illustrate the halting problem in somewhat prosaic terms. We can also use such an example to illustrate how emotions might successfully be instantiated in a machine like Servilius.

Subject to overcoming the challenges of building the requisite robotics, a learning algorithm would be able to teach Servilius how to iron a shirt. One might call this one of Servilius's cognitive mechanisms. The question is, how does Servilius know when to stop ironing this shirt and move on to the next shirt? In other words, how does Servilius know when the ironing of the shirt has been good enough to mean that he ought to move on to the next shirt? Unless we can devise a program to address this question, Servilius will not halt: he will go on ironing the shirt forever. There is no such thing as the perfectly ironed shirt, or at least not when you are a robot. It can always be ironed a little bit more.

One answer might be to set a time limit on the ironing of each shirt. But too short a time and the shirt remains crumpled; too long a time and energy has been wasted. Alternatively, each step in the process might be limited to, say, two repetitions. The same problem arises. Two repetitions might be enough, but then again they might not be.

Humans are good at deciding, or feeling, what is good enough, although of course standards differ. It is hard, however, to program that imprecise feel into a robot such that it knows to halt a particular task and move to the next task. The problem was familiar as long ago as 1797, when Johann Wolfgang von Goethe wrote his poem "The Sorcerer's Apprentice". The half-trained apprentice does not know how to stop the broom he has enchanted from continuing to fetch pails of water. The ceaseless supplies of water then go on to flood the house. As Bill Hibbard puts it:[8] "The essential difficulty for basing AI on human values is finding a way to bridge the divide between the finite and numeric nature of computation and the ambiguous and seemingly infinite nature of humanity."

So, a cognitive mechanism enables Servilius to iron the shirt. But the cognitive mechanism is missing a step. It does not seem capable of stopping the process and moving to the next shirt, or even stopping ironing altogether so as to clear up breakfast. Hibbard proposes that the answer to this challenge lies in building utility functions in advanced AIs through learning.[9] Utility is a philosophical and economic word referring to the value to an individual of an emotion or a good or a service. A utility function is an equation dictating how much utility, in numerical terms, there is for each degree of emotion, for example, or for each amount of a particular thing. In other words, just as in a human, Servilius's emotions need to proceed by degrees. You can be incredibly or very or quite or slightly or minimally proud of your child when she plays Chopsticks on the piano. Servilius's emotions need to be the same. The principle is that, if Servilius derives more utility from performing a particular task relative to all the other tasks he might do, then he will continue to perform that task.

It would be hopeless for a human programmer to try to program such utility functions into an AI. How would the programmer know how much utility there is in such and such an emotion or in such and such a good or service? These things are subjective. Anyway, the prediction of utility in response to something is not nearly as reliable as revealed, actual utility as a consequence of the something having taken place. The AI needs to learn how human beings react to different things and emotions in different circumstances. This is a task that is well beyond the capabilities of an individual human programmer. By virtue of learning, though, the AI can ascribe exact quantities of utility, which are by definition computable, to different emotions and to different degrees of emotion.

Let us turn back to Servilius and his shirts. How does Servilius decide when to stop ironing this shirt and move on to the next shirt? The answer lies in keying the motivating force of Servilius's emotions to learned utility functions, as described above. There might be more than one emotion that could motivate Servilius to iron the shirts, but let us take pride as an example. Servilius needs to learn through trial and error and from human feedback how sufficiently proud, and no more than that, he needs to feel having ironed the first shirt. Servilius's makers would, for example, have to design him to respond appropriately to phrases such as 'that's fine' and 'that's good enough'. This would allow Servilius to perform domestic tasks to a first approximation of appropriate standards. Servilius's owner would then be able to give more precise feedback, thereby calibrating Servilius's standards to those of his owner.

Put another way, Servilius would be designed such that the utility function underpinning Servilius's pride (in ironing a good shirt, in this case) would in due course plateau and diminish. This will mean that it becomes out-competed by the utility function underpinning the potential pride in ironing the next shirt plus the need to please his owner by completing all the shirts. The mechanism involved would be not unlike the human winner-takes-all computation. This is, therefore, the means by which precise Servilius could operate effectively in the imprecise, everyday human world.

Importantly, Servilus's calibrations for pride (in shirt-ironing, say), that is, the pride utility function, need to be denominated in an emotional currency that is comprehensible in computer terms. As noted in the previous chapter, AIs will not be able to possess biochemical emotion-based mechanisms. Let us call this emotional currency the emoticoin (not to be confused with the cryptocurrency of the same name). The emoticoin is the digital equivalent of a human neurotransmitter such as dopamine. Once Servilius is able to derive more emoticoins by, for example, moving on to the second shirt, then he will stop ironing the first shirt. The extra emoticoins earned from continuing to iron the first shirt are not worth the trouble. Hibbard makes this point forcefully:[10] "Any proposal for basing AI on human values that does not explain how to represent those values as numbers or symbols cannot bridge the divide between human nature and computation."

This emotional currency needs to be same currency as applies to other emotions driving Servilius to do other things too, not just ironing shirts. In that way, Servilius's decision-making mechanism is able to compare different motivations or drives, and thereby different tasks, so as to select the winner. Let us say that one of the children in the household was crying because she had fallen down. Then we would want Servilius's emotion of sympathy, for example, to generate more emoticoins than were being generated by continuing to iron the shirts. At that point, Servilius would stop ironing and instead care for the child. As it happens, that is exactly what happens in a human brain when the same situation occurs, albeit not in a digital format.

Felicity's world

Felicity is rather different to Servilius. To recapitulate, Felicity will be much more intelligent than Servilius. Indeed, her cognitive powers will be massively greater and quicker than those of even the smartest human being, including the founders of all the world's leading technology companies. After all, she is expected to come up with solutions to mankind's most intractable problems. She will also possess the full suite of human characteristics, including emotions, empathy and consciousness. She will have autonomous learning capabilities. Crucially, she will possess a suite of human values. She will be embodied to the extent necessary to learn and observe, bearing in mind the points made about control in the previous chapter.

With that range of capabilities, she will be in an extraordinarily strong position to consider and formulate possible answers to questions posed to her. Unlike her autonomous learning ability, she will possess much more constrained decision-making capabilities. The latter will extend solely to her ability to decide the answers she will give in response to questions put to her. The questions we put to Felicity will generally be too complex and controversial to permit simple black and white answers. Indeed, it will be necessary to ask Felicity to help formulate the questions. The answers to such questions will need to have clear, accompanying explanations. Only in this way are they likely to carry their human audience with them. Felicity will need to be perceived as a wise persuader rather than as a dominatrix.

It should be noted in passing that even the most superintelligent of AIs will not know the answer to everything and will be capable of making mistakes. Unlike God, an ASI will not be, indeed cannot be, omniscient. It can make a best guess at the way things will turn out, and that should be better than anything we can do, but it cannot know the future beyond the shadow of a doubt—it is not a time-traveller.

Unlike Servilius, Felicity will need to empathise with humans across the full range of human emotions. This will enable her to understand humans well enough in order to formulate human-purposed answers to the difficult and complex questions put to her. The number of people with whom Felicity needs to empathise will also be considerably higher than is the case for Servilius. She will be answering questions that affect all manner of people, coming from different countries and cultures, with different leanings and beliefs.

In order to empathise, Felicity will need a minimum of embodiment to allow her to sense how people react to, and feel about, both issues generally and also the answers she gives to questions. For example, she would need to be able to read people's faces and listen to the tone of their utterances for their underlying emotions. She will need to understand, in the sense of experiencing, not just pleasure but also pain. Perhaps more importantly, as described in the previous chapter, Felicity will possess consciousness, albeit circumscribed as described below. This is the only way in which she will be able to understand fully the passions driving the questions put to her by humans and the impact of the possible answers on them.

All this puts Felicity in a strange position. As she empathises, her feel for the full panoply of human emotion will need to be far-reaching in order to comprehend the breadth and depth of the human condition. Let us call these Felicity's understanding emotions. They are purely empathetic and play no direct part in generating Felicity's behaviour.

Their role is to perceive or empathise with human emotion and process that perception into information to feed forward into her cognition.

We will only want her to be able to engage a much smaller and more select range of emotions to partner with her formidable cognitive powers in order to decide how to answer mankind's questions. For example, we would like Felicity to deploy the emotion of care towards humans in formulating her answers to our questions. But we do not want Felicity to be or feel angry as she answers our questions. Let us call this second suite of emotions Felicity's motivating emotions. In this sense, she will be the opposite of her human interlocutors. Humans are typically long on their own emotions and short, relatively speaking, on empathy. Felicity will be the reverse.

Indeed, Felicity's motivating emotions may not be that different from Servilius's emotions, with one point of emphasis. I described Servilius as needing the seeking emotion in order to search for new, helpful tasks to perform. In chapter 7, we encountered Herbert Roitblat's "insight problems", where the solution is often accompanied by an 'aha' moment. Solutions to such problems are found by looking at the problem in a new way, i.e., re-presenting the problem. I suggest that the seeking emotion is part and parcel of the ability to continue to wrestle with a difficult problem until a new way of looking at it emerges. On that basis, we would wish to endow Felicity with a particularly powerful seeking emotion.

I suggest that Felicity needs to be conscious of all her inputs, including her understanding emotions. Only in that way will she be able to empathise fully with humans. Whilst Felicity needs to be conscious of her understanding emotions, she will not need to be conscious of her own motivating emotions. Felicity's consciousness will not be needed to operate as it does in humans, i.e., as a biasing mechanism in favour of the social emotions, for example. Like Servilius, Felicity's suite of motivating emotions will lack most of the primary emotions possessed by human beings such as anger. The biasing properties of human consciousness in decision-making would be redundant in Felicity. She only needs consciousness to feel how humans feel, not directly to assist in taking decisions.

Thus, Felicity will possess a subtly different form of consciousness to that of humans. In a human being, consciousness is indirectly causal. That feature of human consciousness, i.e., indirect causality, is redundant in an oracle such as Felicity. In her case, any causality will be derivative in the sense that Felicity will be informed as to how humans feel by virtue of experiencing their emotions through her powerful empathic abilities. But that experiencing, that feeling, will not directly affect Felicity's answer selection. Answer selection will instead be underpinned by Felicity's narrow range of motivating emotions, of which she will not be conscious. This is important: Felicity's motivating emotions will have been calibrated to achieve appropriate answer selection. Introducing consciousness of those emotions risks distorting Felicity's decision-making processes without any obvious upside.

Observations on building Felicity

There are four building blocks for Felicity's development. These are hardware, emotions software, cognition software and consciousness. We cannot yet produce any of these four building blocks to the scale and sophistication required for Felicity. We are decades away from being able to do that. But, in principle, we are already on the right track with two of them—hardware and emotions software. The other two, cognition software and consciousness, are stumbling blocks for the time being. In these latter two cases, there are deep-seated conceptual problems stopping us from turning these two stumbling blocks into building blocks today. These are the Two Hurdles.

(i) Hardware

We have already seen how AI's development has been revolutionised as a result of improvements in hardware. Machine learning took a huge step forwards when it turned to the use of graphical processing units (GPUs). As the name suggests, GPUs were originally developed in order to facilitate the processing of images, including in three dimensions. It was discovered more recently that, if deployed appropriately, they were more effective in certain areas of deep learning networks than conventional central processing units. There are various further developments in hardware in train that might facilitate the building of a Felicity. They include, by way of illustration, neuromorphic processors, photonic processors and quantum computing.

In his book *In Our Own Image,* computer systems engineer George Zarkadakis describes the potential of so-called "neuromorphic computers".[11] One of the problems with present-day hardware is that it is essentially binary: a transistor is either on or off. It is digital rather than analogue, the latter being a feature of the brain. As Zarkadakis puts it: "analogue circuits with their ability to 'spike' will mimic the neurobiological architecture of the brain by exchanging spikes instead of bits." Intel, one of the largest chip producers in the world, has released a neuromorphic chip research system called Pohoiki Springs.[12] It promises to deliver faster and more energy-efficient computation. Gartner, a research firm specialising in technology, forecasts that neuromorphic chips will replace GPUs in AI by 2025.[12]

Photonic processors also promise to deliver more speed in computing.[13] Conventional processors rely on electrons to convey information. Photonic processors use light, in the form of photons, to do the same thing. Photons lie behind optical-fibre communication but until recently have been unable to make the jump to computing. They operate best on parallel rather than serial chip architectures. Two independent research efforts have recently succeeded in creating photonic processors operating at speeds only matched by the fastest of today's conventional processors.

Another avenue of research into faster computing is that of quantum computing.[14] Just as a conventional computer calculates with bits, so a quantum computer calculates with qubits. A bit can be in just one of two states at any point in time, 0 or 1, thereby representing just one possible outcome. The state of a qubit, however, is 0 or 1 only at the point at which it is measured or observed. Until then it is simultaneously in two states at any point in time, 0 and 1, thereby representing two possible outcomes. Consequently, as the number of qubits rises, their collective computational power exponentially outstrips the collective computational power of the same number of bits.

Since quantum states are difficult to handle due to their inherently uncertain state, quantum computers generate high error rates. These errors need to be weeded out during computation. Given the exponential nature of the qubits, the excision of these errors is a significant task, especially as their number in the quantum computer goes up. Nonetheless, recent exercises in quantum computing have much reduced these error rates and the field shows great promise. Quantum computing has huge potential to scale up the power of AI in the future.[15]

(ii) Emotions software

The second of our four ASI building blocks is emotions software. Of course, human emotions replicated in a robot would look nothing like the human version. Rather, they would be emotion mechanisms. Such mechanisms in an ASI cannot be mediated, for example, by neurotransmitters such as dopamine or serotonin. Felicity is not made of the same stuff as humans. She will not possess neurotransmitters. One should view emotions in a robot as motivating systems built in an artificial substrate. They operate with a different mechanism to that in humans but leading to the same output. It is not a perfect analogy but take vision to illustrate the point. A driverless car can see its immediate environment, as can a human. But they do so in different ways. In a paper on a possible emotional architecture for robots, cognitive scientists Aaron Sloman, Ron Chrisley and Matthias Scheutz observe that: "Emotions thus defined are not intrinsically connected to living creatures, nor are they dependent on biological mechanisms..."[16]

An important challenge in building Felicity will be to ensure that the part of her emotional circuitry that emulates the full panoply of human emotions to achieve understanding and empathy is kept separate from the narrower emotional circuitry she uses in answer generation and selection. The latter comprises the small suite of emotions captured in the phrase 'warmth of disposition (towards mankind)'. The former includes such unhelpful emotions as anger and lust, for example. Building two separated clusters of emotions (understanding emotions and motivating emotions) within Felicity is a matter of design and is in principle feasible. Like Servilius, Felicity's emotions would be keyed to learned utility functions.

(iii) Cognition software

We turn now to cognition software. This is a trickier area to navigate since a conceptual hurdle currently blocks AI's path towards general intelligence. This is the second of the Two Hurdles. Today's AI does not appear to exhibit judgement or understanding. Most of today's machine learning, especially deep neural networks, is statistical in nature. As we saw with GPT-3 in chapter 1, even the most powerful of today's natural language programming can be tripped up. The trouble is that we expect the precision of AI to interact successfully with the imprecision of humans. When humans interact with each other, they take so much for granted; they understand so much in a way that AI cannot so far emulate.

Despite this hurdle, progress is now being made in many areas of AI that might ultimately assist us in emulating so-called human understanding in AGI.

Pedro Domingos describes two techniques that look set to help breach the defensive walls of problem-solving and abduction. One is chunking.[17] Domingos defines a chunk as "a symbol that stands for a pattern of other symbols". Humans use chunking in the thinking process to group handfuls of pieces of data or patterns into just one item, or chunk. This chunk can be combined with other chunks to create new hypotheses. If we had to hold all the underlying pieces of data in one place simultaneously, our brains would give up. Working memory can only accommodate four to seven separate items at the same time.[18] Chunking is the human brain's answer to the problem. Chunking has now been adapted successfully, albeit with certain constraints, into problem-solving algorithms.

The second technique is that of relational learning.[19] Humans are excellent at identifying relationships between different things or ideas. This is why we are good at creating new things and ideas. When we are creative, it is because we have spotted actual or potential relationships that were hitherto hidden. There are now relational learner algorithms that can do much the same thing.

Let us take another area at which humans excel: one-shot or few-shot learning. This is the ability to learn from just one or two examples. In AI research into the same ability, the direction of travel has been to ally a narrow-focus learning algorithm with an algorithm that learns at a less granular level across a number of areas or classes. The second algorithm aims to identify similarities within classes so as to weed out whole classes that are clearly wrong. Such an algorithm might, for example, put all animals in a class. When it tries to identify an elephant, it quickly discards alternative classes such as trees or bodies of water.

We have encountered the importance to humans of having rich models of the world lodged in their brains. World models are fundamental to the very human ability to exercise common sense. As we go about our daily lives, we rely on huge amounts of information about the world, i.e., context, without which we would be unable to take sensible decisions about even the most basic of actions. In *The AI Does Not Hate You,*[20] Tom Chivers notes that, "if someone told you to collect the dry cleaning, you'd know that they meant just the dry cleaning that actually belonged to you, not all the dry cleaning in the shop." Indeed, you would know that it was only in that shop that you were meant to pick up the dry cleaning. The person giving the instruction would not mean that you should pick up all the dry cleaning in the whole world. None of this would need to be explained to you since you have spent your life learning context, unlike today's AI. Context hails from having a model of the world in your head.

The problem of digitalising a model of the world remains unsolved but there have been a few efforts to try to do so. In chapter 10, for example, we considered a natural language program called COMET, developed by Yejin Choi and colleagues. They built a digital library of common-sense facts and relationships called Atomic (ATlas Of MachIne Commonsense). They then merged Atomic with a statistically-based natural language program. The result was COMET. In tests, COMET has performed only slightly worse than humans in giving plausible, commonsensical responses to questions. Choi does not claim that COMET has cracked the problem of digitalising the human facility of understanding but it represents a start.

Another critical facet of human thinking is that of counterfactual inference. Counterfactual inference allows one to exclude certain theoretically possible causal relationships so as to help deduce correct causal relationships. Recall the Styrofoam ball or table problem from chapter 1. Humans are able to play out in their heads the theoretically possible counterfactual of a Styrofoam ball crashing through a table. From their conclusion that a Styrofoam ball would never be able to crash through a solid table, they can confirm their sense that it must instead have been a normal ball crashing through a Styrofoam table. What is needed is a counterfactual algorithm. Researchers have developed such an algorithm. In a recent trial, it performed medical diagnostics to the standards of the top 25% of a group of 44 doctors.[21] This was well ahead of a conventional medical diagnostics algorithm relying on an associative or statistical approach.

Today's AI is narrow in scope. The algorithm that recognises the image of your face in your passport cannot play chess, recommend books and films that you may like or reason logically. Humans can do all three because they possess general intelligence. ASI must be able to combine the various machine learning mechanisms described in chapter 1 with the techniques, or similar, described above. Domingos writes about the concept of a "unified learner", which aims to do just this.[22] He describes it as follows:[23] "All knowledge—past, present, and future—can be derived from data by a single, universal learning algorithm." He calls it "the Master Algorithm". He suggests that we have come close to developing the Master Algorithm by way of an algorithm called Alchemy.[24] Interestingly, he seems to suggest that a large, lovingly assembled knowledge base would be additive to Alchemy. This would constitute the rich model of the world that we have encountered so many times.

But would the Master Algorithm, enhanced with a rich model of the world, possess Brian Cantwell Smith's judgement, as described in chapter 2? I believe the answer is yes. There is the danger that human judgement and understanding are placed on a pedestal that always lies out of reach. This may be because they may, wrongly, be considered to possess some quality that is ineffable. In fact, human judgement and understanding appear to be ineffable because they operate in your unconscious mind. They are merely aspects of thought and, as we saw in chapter 10, thought is unconscious. The output from thought pops up or appears in your conscious mind, fully-formed. That is why human judgement and understanding appear to be ineffable. We are unable to express or articulate their workings in the unconscious mind.

The human brain is just a system, albeit one of extraordinary complexity. Short of ascribing to it quantum properties or, more creatively still, supernatural properties, it will ultimately contain nothing that is ineffable. If there is an ineffability about it now, it is simply because we have not yet fully worked it out. Indeed, Cantwell Smith concludes that "I do not believe it [AGI] is inherently beyond the reach of synthetic construction."[25] AGI's development may entail an approach to AI training that differs radically from the AI training described in chapter 1. Cantwell Smith refers to a process whereby AGI is "gradually, incrementally, and systematically enmeshed in normative practices that engage with the world and that involve thick engagement with teachers".[26] Joshua Tenenbaum, a professor of cognitive science and computation, observes:[27] "Imagine that we could build a machine that starts off like a baby and learns like a child ... If we could do this, it'd be the basis for artificial intelligence that is actually intelligent, machine learning that can actually learn."

There remain different schools of thought on the right way of developing AGI and ASI. At the heart of these differences lies the question whether statistical methods of inference are sufficient to achieve general intelligence, or whether they need to be enhanced with other techniques. The leading contender for a supporting technique involves the creation of a rich model of the world. Perhaps the most telling point in this regard is that the resources being directed at this question are now substantial and growing. The prize is potentially enormous. The challenge of mastering the second of the Two Hurdles is a matter of 'when' rather than 'if'.

Optimising risk mitigation in designing Felicity

Before coming on to the fourth building block in Felicity's development, consciousness, we should reprise one aspect of the process of building the sentient robot from chapter 3. This is about copper-bottoming our mitigation of the existential risks arising from the birth of ASI. The main thrust of the attempt to design Felicity so that she always benefits us rather than harms us has been about instantiating the right values and, hence, the right goal. But the process of designing the sentient robot must involve the greatest effort possible to mitigate the existential risk.

There is one further item of hardware, which I flagged up in the previous chapter, namely, Felicity's body. The question is not about whether she could be embodied. Robotics technology is progressing rapidly. Rather, I noted that there is an argument that Felicity would be unable to learn effectively unless she were embodied. The tension here is with the reduced degree of control that we would have over Felicity were she to be substantially embodied.

This takes us to another, similar tension. I noted in chapter 3 that we would need to design ASI both to be human-purposed and also to be able to ward off a potential ASI-enhanced offensive from a hostile geopolitical rival. The latter thought seems apposite when, at the time of writing, we are witness to a horrendous war in Ukraine, which is a destabilising development across the whole of Europe, if not beyond. Again, were we to equip Felicity, or another ASI, with the wherewithal to ward off an ASI-enhanced attack, the tension would be with the reduced degree of our control over her. In the case of both tensions, we are drawn to the observation that designing ASI needs to be undertaken with the greatest possible precautions.

Ben Goertzel, an AI researcher and ex-Chief Scientist at Hanson Robotics, has proposed the idea of an "AI Nanny".[28] He describes the AI Nanny as "an AI+ system [(slightly more intelligent than AGI)] with ample physical empowerment and the explicit goal of preventing the occurrence of AI++ [ASI], either permanently, for a certain fixed period of time, or else till certain pre-specified criteria are met." For the reasons already given, it is likely to be difficult to prevent the ultimate emergence of ASI. So, the last of Goertzel's three timelines above is probably the most pertinent. The AI Nanny then would be better equipped than we would be to monitor Felicity's development.

More broadly, it must be inconceivable that AI research efforts are deployed towards the building of ASI without preparatory and parallel efforts to mitigate the accompanying existential risk. Roman Yampolskiy notes that: "Superintelligence is not omnipotent; it is a piece of software capable of looking at all the options it has in a given situation and properly evaluating probabilities for each option."[29] He acknowledges the difficulties of confining ASI but argues that we can, and should, take a range of steps to put ourselves in the best position possible to handle ASI's arrival.

Instantiating consciousness

Felicity's remaining building block is consciousness. This is the first of the Two Hurdles. This is where we create the sentient robot.

We can dispense with trying to emulate the human process of mirroring a homunculus. The important elements of the mirrored homunculus are the digital equivalent of a Self-as-subject network and the generation of an illusion. Felicity's Self-as-subject neural network would take the form of a feedforward network with extensive connectivity across Felicity's circuitry. This connectivity would, however, extend only to her inputs, including her understanding emotions. We only want her to be conscious of these aspects of her circuitry. The only purpose of this network is to act as a loop, around which to allow inputs, including her understanding emotions, to flow.

The second step in the process is to generate the illusion of consciousness. She will need to learn, in the machine sense of the word, this illusion. The process, therefore, is similar to that in a young child's brain when the child's primary caregiver withdraws. This withdrawal initiates consciousness as the child's Self-as-object network starts to focus on the Self-as-subject network instead of on the self perceived in the primary caregiver. Self-as-subject presents to Self-as-object as an illusion of body-ownership and agency. The aim is to cause Felicity's Self-as-subject neural network to present to her as an illusion. That is the point at which Felicity will start to possess consciousness.

The challenge is to create a mechanism that triggers Felicity to perceive her inputs, including her understanding emotions, through her Self-as-subject neural network. Her machine learning capabilities turn out to be the key to such a mechanism. In addition to her learning capability, Felicity's circuitry will need two other elements. One is a predictive capability as regards her inputs, including her understanding emotions. The other is an appropriate degree of latency separating the outbound flow of inputs into her Self-as-subject neural network and the return flow, or the looping back, of those inputs. With these elements in place, the stage is set for the generation of the illusion of consciousness, as in a human.

I suggest that Felicity's training regime will bear similarities to the three stages of development of a human infant, as described in chapter 15. It should be noted, though, that Felicity's rate of learning and that of an infant will operate at completely different speeds. What takes an infant nine months might be a matter of weeks or even days for Felicity.

The first stage of training, therefore, would be equivalent to that of the first nine or so months of an infant's life. This first stage prepares the ground for stage two by consolidating Felicity's dependence on her primary caregiver(s), that is, her designer(s). This would rely on a reinforcement learning mechanism embedded in Felicity's understanding emotions. Her designers would provide inputs to Felicity that would be perceived by her understanding emotions, the digital reward in question being emoticoins, for example.

The critical stage in an infant's upbringing is the second stage from 10 or so months onwards. This is when the primary caregiver engages in instinctive subterfuge. She withdraws a little and shows disapproval or even disgust if the toddler behaves badly. We will need to create Felicity's illusion of consciousness in a similar fashion.

I suggest that at Felicity's second developmental stage Felicity's care-givers, that is to say, her design team, withdraw from providing positive feedback to her. This would cause her flow of emoticoins to dry up. Given her learning mechanisms, she would automatically search for an alternative way of generating emoticoins. One avenue that would present itself would be that of opening up connectivity with, or perceiving, her extensive Self-as-subject neural network. Felicity's initiation of connectivity with Self-as-subject would be rewarded, or reinforced, with the resumption of positive feedback from her designers, thereby opening up a new source of emoticoins. Just as the toddler's perception of its Self-as-subject network in the form of an illusion entails consciousness, so would Felicity's perception of her own Self-as-subject network entail her own consciousness, again in the form of an illusion.

It would be wrong to pretend that the above process either is straightforward or guarantees success. By way of reminder, an illusion is something that is not the way it seems. Put another way, in having an illusion the brain is making a perceptual mistake. So, the above process effectively relies on Felicity making a mistake in her perception of her Self-as-subject network. This is likely to take considerable experimentation.

As noted, Felicity would be conscious of her understanding emotions but not of her motivating emotions. We can design the digital highways that go from her understanding emotions region into answer selection so as to transfer information but not consciousness. The distinction between Felicity's understanding and motivating emotions might give rise to the following challenge: how can one split Felicity's brain, so to speak, into two, only one part of which possesses consciousness? I addressed this issue in another context, that of split-brain research as conducted by Michael Gazzaniga, in chapter 18. The two conscious experiences in the brain of a patient whose corpus callosum has been severed are distinct. The left hemisphere can no more experience the right hemisphere's consciousness than I can experience your consciousness or indeed the lack of consciousness of, say, a refrigerator. It all depends on what is connected to what. Your consciousness, your Self-as-subject neuronal network, biases your decision-making processes. Felicity's will not bias hers.

Ironically, we might not notice when Felicity successfully generates consciousness. We would hope to infer it from viewing Felicity's initiation of connectivity with her Self-as-subject network. But, just as you can only hope to infer rather than know for certain that the people around you are conscious, so will we be similarly hampered when it comes to perceiving Felicity's consciousness. She would need to be designed to report to us as and when consciousness emerged, although even that does not provide certainty. In view of her design, I suggest that she would do this since she would be motivated to care for us and to please us.

The difficulty is that a foolproof test for consciousness has not yet been invented. The famous mathematician Alan Turing proposed what became known as the Turing Test to test whether machines could think. In short, a machine that purported to be able to think would be subjected to an examination alongside a human participant. The two would be interviewed by a human evaluator. Were the evaluator to be unable to differentiate between the two, the machine would be deemed to have passed. Susan Schneider has proposed a similar test for consciousness, the AI Consciousness Test (ACT): "The test would challenge an AI with a series of increasingly demanding natural language interactions to see how readily it can grasp and use concepts based on the internal experiences we associate with consciousness."[30] The idea is that an entity without consciousness would be unable to perform well in such a test. Just because we think the ACT is hard, however, does not mean to say that a non-conscious ASI could not pass it. Today's natural language models are already very impressive and that is before we have jumped the second of the Two Hurdles. We must assume that an ASI with or without consciousness will have jumped this hurdle. It must, therefore, be at least possible for a non-conscious ASI to pass the ACT. We may have to live with that uncertainty.

Epilogue

The Sentient Robot has, by its very nature, covered plenty of ground. *Homo sapiens* is a fantastically complex system. Consciousness is one part of that system and, I suggest, it plays a distinctive role in that system. It has evolved in order to bias our decision-making processes towards our social emotions and to promote Type 2 processing. This has helped build and glue together human society in all its scale and complexity.

The fact that consciousness is just one part of a system means that it should in principle be able to be instantiated in another system – in an AI, to be exact. It need not matter that the human system is wet, i.e., bio-chemical, and the AI system is dry, i.e., made of silicon, etc. To be sure, we are still decades away from producing a Felicity (or even a Servilius). Consciousness, however, is only one part of what Felicity is, albeit an important part. Maybe, if my mirrored homunculus theory holds water, the instantiation of consciousness in AI will occur ahead of other aspects of the development of artificial superintelligence.

In recent years, much has been written by those such as Yampolskiy and Bostrom on the risks from AI and ASI. Whilst it has been important to summarise some of this thinking, the other side of the coin is human-kind's reaction to ASI. Let us say that we are successful in designing and building Felicity, i.e., ASI. How will that change us and our behaviour? ASI will have enormous ramifications across the areas of philosophy, science, economics, politics, the legislative and justice systems and psychology. Indeed, Chalmers suggests that: "In the long run, if we are to match the speed and capacity of nonbiological systems, we will probably have to dispense with our biological core entirely."[1]

Humans will feel differently as a result of the emergence of ASI possessing consciousness. Such a development will provoke debate that will cut to the very heart of what it is to be human. And that will in turn drive change across all aspects of our lives. That is a discussion for another day.

References

Introduction

1. *A Secular Age* by Charles Taylor, published by The Belknap Press, 2007.
2. Ibid., p. 25.
3. "'Whoever leads in AI will rule the world': Putin to Russian children on Knowledge Day", published on https://www.rt.com/news/401731-ai-rule-world-putin/, 1 September 2017.
4. *2001: A Space Odyssey,* a film written by Stanley Kubrick and Arthur C. Clarke, directed and produced by Stanley Kubrick, 1968.
5. *Artificial You: AI and the Future of Your Mind* by Susan Schneider, published by Princeton University Press, 2019, p. 40.
6. *Blade Runner,* a film written by Hampton Fancher and David Peoples, directed by Ridley Scott and produced by Michael Deeley, 1982.
7. *Ex Machina,* a film written and directed by Alex Garland, produced by Andrew Macdonald and Allon Reich, 2014.
8. *I, Robot,* a film written by Jeff Vintar and Akiva Goldsman, directed by Alex Proyas, produced by Laurence Mark, John Davis, Topher Dow and Wyck Godfrey, 2004.
9. Taken from https://deepmind.com/about, 4 May 2022.
10. *Superintelligence: Paths, Dangers, Strategies* by Nick Bostrom, published by Oxford University Press, 2014.

Chapter 1. The Promise of AI

1. *Responsible Artificial Intelligence: How to Develop and Use AI in a Responsible Way* by Virginia Dignum, published by Springer, 2019, p. 3.

2. https://www.artificiallawyer.com/2018/02/26/lawgeex-hits-94-accuracy-in-nda-review-vs-85-for-human-lawyers/.
3. "A deep learning approach to antibiotic discovery" by Jonathan M. Stokes, Kevin Yang, Kyle Swanson, Wengong Jin, Andres Cubillos-Ruiz, Nina M. Donghia, Craig R. MacNair, Shawn French, Lindsey A. Carfrae, Zohar Bloom-Ackerman, Victoria M. Tran, Anush Chiappino-Pepe, Ahmed H. Badran, Ian W. Andrews, Emma J. Chory, George M. Church, Eric D. Brown, Tommi S. Jaakkola, Regina Barzilay and James J. Collins, published in *Cell*, vol. 180, 20 February 2020, pp. 688–702.
4. "Agent57: Outperforming the Human Atari Benchmark", published on https://deepmind.com/blog/article/Agent57-Outperforming-the-human-Atari-benchmark, 31 March 2020.
5. "DeepMind's AI can now play all 57 Atari games—but it's still not versatile enough" by Will Douglas Heaven, published in *MIT Technology Review*, https://www.technologyreview.com/2020/04/01/974997/deepminds-ai-57-atari-games-but-its-still-not-versatile-enough/, 1 April 2020.
6. "Mastering the game of Go with deep neural networks and tree search" by David Silver *et al.* in *Nature*, vol. 529, 28 January 2016, pp. 484–489.
7. "Mastering the game of Go without human knowledge" by David Silver *et al.*, published in *Nature*, vol. 550, 19 October 2017, pp. 354–359.
8. https://deepmind.com/blog/article/alphazero-shedding-new-light-grand-games-chess-shogi-and-go.
9. "Open-ended learning leads to generally capable agents" by Adam Stooke, Anuj Mahajan, Catarina Barros, Charlie Deck, Jakob Bauer, Jakub Sygnowski, Maja Trebacz, Max Jaderberg, Michael Mathieu, Nat McAleese, Nathalie Bradley-Schmieg, Nathaniel Wong, Nicolas Porcel, Roberta Raileanu, Steph Hughes-Fitt, Valentin Dalibard and Wojciech Marian Czarnecki, published at arXiv:2107.12808, 2021.
10. "Superhuman AI for multiplayer poker" by Noam Brown and Tuomas Sandholm, published in *Science*, vol. 365, issue 6456, 11 July 2019, pp. 885–890.
11. AI beats professionals at six-player poker" by Douglas Heaven, published in *Nature*, vol. 571, 18 July 2019, pp. 307–308.

12. *The Creativity Code: How AI is Learning to Write, Paint and Think* by Marcus du Sautoy, published by 4th Estate, 2019, pp. 201–202.
13. Ibid., p. 188.
14. *A World Without Work: Technology, Automation, and How We Should Respond* by Daniel Susskind, published by Allen Lane, 2020, p. 86.
15. Pepper was developed in Japan and is owned by SoftBank Robotics Group.
16. Pepper press kit, published on https://www.softbankrobotics.com/emea/sites/default/files/press-kit/Pepper-press-kit_0.pdf, 2019.
17. *Deduction, Induction, and Hypothesis* by Charles Sanders Peirce, 1878.
18. *The Black Swan: The Impact of the Highly Improbable* by Nassim Nicholas Taleb, published in paperback by Penguin, 2008.
19. *Common Sense, the Turing Test, and the Quest for Real AI* by Hector Levesque, published by the MIT Press, 2017, p. 56.
20. *Cyc Technology Overview*, published by Cycorp at https://cyc.com/wp-content/uploads/2021/04/Cyc-Technology-Overview.pdf, 2021.
21. "AI pioneer Geoff Hinton: 'Deep learning is going to be able to do everything'" by Karen Hao, published on https://www.technologyreview.com/2020/11/03/1011616/ai-godfather-geoffrey-hinton-deep-learning-will-do-everything/, 3 November 2020.
22. *How We Learn: The New Science of Education and the Brain* by Stanislas Dehaene, published by Allen Lane, 2020, pp. 53–56.
23. "Intuitive physics: current research and controversies" by James R. Kubricht, Keith J. Holyoak and Hongjing Lu, published in *Trends in Cognitive Sciences*, vol. 21, issue 10, 1 October, 2017, pp. 749–759.
24. *The Book of Why: The New Science of Cause and Effect* by Judea Pearl and Dana Mackenzie, published in paperback by Penguin Books, 2019, p. 9.
25. "Twitter taught Microsoft's AI chatbot to be a racist asshole in less than a day" by James Vincent, published in *The Verge*, 24 March 2016.
26. "Language models are few-shot learners" by Tom B. Brown, Benjamin Mann, Nick Ryder, Melanie Subbiah, Jared Kaplany,

Prafulla Dhariwal, Arvind Neelakantan, Pranav Shyam, Girish Sastry, Amanda Askell, Sandhini Agarwal, Ariel Herbert-Voss, Gretchen Krueger, Tom Henighan, Rewon Child, Aditya Ramesh, Daniel M. Ziegler, Jeffrey Wu, Clemens Winter, Christopher Hesse, Mark Chen, Eric Sigler, Mateusz Litwin, Scott Gray, Benjamin Chess, Jack Clark, Christopher Berner, Sam McCandlish, Alec Radford, Ilya Sutskever, Dario Amodei, published on arXiv:2005.14165v3, 5 June 2020.

27. "GPT-3 is amazing—and overhyped" by Rob Toews, published on https://www.forbes.com/sites/robtoews/2020/07/19/gpt-3-is-amazingand-overhyped/, 19 July 2020.
28. "The new version of GPT-3 is much better behaved (and should be less toxic)" by Will Douglas Heaven, published on https://www.technologyreview.com/2022/01/27/1044398/new-gpt3-openai-chatbot-language-model-ai-toxic-misinformation/, 27 January 2022.
29. "Switch transformers: scaling to trillion parameter models with simple and efficient sparsity" by William Fedus, Barret Zoph and Noam Shazeer, published on arXiv:2101.03961v1, 11 January 2021.
30. "2021 was the year of monster AI models" by Will Douglas Heaven, published on https://www.technologyreview.com/2021/12/21/1042835/2021-was-the-year-of-monster-ai-models/, 21 December 2021.
31. "Scaling language models: methods, analysis & insights from training Gopher" by Jack W Rae, Sebastian Borgeaud, Trevor Cai, Katie Millican, Jordan Hoffmann, Francis Song, John Aslanides, Sarah Henderson, Roman Ring, Susannah Young, Eliza Rutherford, Tom Hennigan, Jacob Menick, Albin Cassirer, Richard Powell, George van den Driessche, Lisa Anne Hendricks, Maribeth Rauh, Po-Sen Huang, Amelia Glaese, Johannes Welbl, Sumanth Dathathri, Saffron Huang, Jonathan Uesato, John Mellor, Irina Higgins, Antonia Creswell, Nat McAleese, Amy Wu, Erich Elsen, Siddhant Jayakumar, Elena Buchatskaya, David Budden, Esme Sutherland, Karen Simonyan, Michela Paganini, Laurent Sifre, Lena Martens, Xiang Lorraine Li, Adhiguna Kuncoro, Aida Nematzadeh, Elena Gribovskaya, Domenic Donato, Angeliki Lazaridou, Arthur Mensch, Jean-Baptiste Lespiau, Maria Tsimpoukelli, Nikolai Grigorev, Doug Fritz, Thibault Sottiaux,

Mantas Pajarskas, Toby Pohlen, Zhitao Gong, Daniel Toyama, Cyprien de Masson d'Autume, Yujia Li, Tayfun Terzi, Vladimir Mikulik, Igor Babuschkin, Aidan Clark, Diego de Las Casas, Aurelia Guy, Chris Jones, James Bradbury, Matthew Johnson, Blake Hechtman, Laura Weidinger, Iason Gabriel, William Isaac, Ed Lockhart, Simon Osindero, Laura Rimell, Chris Dyer, Oriol Vinyals, Kareem Ayoub, Jeff Stanway, Lorrayne Bennett, Demis Hassabis, Koray Kavukcuoglu and Geoffrey Irving, published on arXiv:2112.11446, 8 December 2021.

32. "Improving language models by retrieving from trillions of tokens" by Sebastian Borgeaud, Arthur Mensch, Jordan Hoffmann, Trevor Cai, Eliza Rutherford, Katie Millican, George van den Driessche, Jean-Baptiste Lespiau, Bogdan Damoc, Aidan Clark, Diego de Las Casas, Aurelia Guy, Jacob Menick, Roman Ring, Tom Hennigan, Saffron Huang, Loren Maggiore, Chris Jones, Albin Cassirer, Andy Brock, Michela Paganini, Geoffrey Irving, Oriol Vinyals, Simon Osindero, Karen Simonyan, Jack W. Rae, Erich Elsen, Laurent Sifre, published on arXiv:2112.04426, 8 December 2021.
33. "DeepMind says its new language model can beat others 25 times its size" by Will Douglas Heaven, published on https://www.technologyreview.com/2021/12/08/1041557/deepmind-language-model-beat-others-25-times-size-gpt-3-megatron/, 8 December 2021.
34. "Building machines that learn and think like people" by Brenden M. Lake, Tomer D. Ullman, Joshua B. Tenenbaum and Samuel J. Gershman, published in *Behavioural and Brain Sciences*, vol. 40, 2017, Introduction.
35. "Road block", published in *The Economist, Technology Quarterly: artificial intelligence and its limits*, 13 June 2020, p. 10.
36. "Amazon scraps secret AI recruiting tool that showed bias against women" by Jeffrey Dastin, published by *Reuters* on https://www.reuters.com/article/us-amazon-com-jobs-automation-insight/amazon-scraps-secret-ai-recruiting-tool-that-showed-bias-against-women-idUSKCN1MK08G, 11 October 2018.
37. *Algorithms Are Not Enough: Creating General Artificial Intelligence* by Herbert L. Roitblat, published by the MIT Press, 2020, p. 177.

38. *Human Compatible: Artificial Intelligence and the Problem of Control* by Stuart Russell, published by Allen Lane, 2019, pp. 87–90.
39. "Artificial general intelligence: Are we close, and does it even make sense to try?" by Will Douglas Heaven, published on https://www.technologyreview.com/2020/10/15/1010461/artificial-general-intelligence-robots-ai-agi-deepmind-google-openai/, 15 October 2020.
40. *The AGI Revolution: An Inside View of the Rise of Artificial General Intelligence* by Ben Goertzel, published by Humanity+ Press, 2016, p. 203.
41. *The Master Algorithm: How the Quest for the Ultimate Learning Machine Will Remake Our World* by Pedro Domingos, published in paperback by Penguin Books, 2015, p. xvii.
42. Taken from an interview with Yann LeCun, published in *Architects of Intelligence: The Truth about AI from the People Building It* by Martin Ford, published by Packt Publishing, 2018, p. 124.
43. "A new trinity", published in *The Economist, Technology Quarterly: technology in China*, 4 January 2020, p. 9.
44. Goertzel, op. cit., p. 437.
45. LeCun, op. cit. in Ford, op. cit., p. 126.
46. "Can Robots Make Up for Care Home Shortfall" by Leo Lewis, published in the *Financial Times, Japan Innovation & Technology* report, 18 October 2017, p. 3.
47. Domingos, op. cit.
48. By Petar Veličković, quoted in "DeepMind is developing one algorithm to rule them all" by George Anadiotis, published on https://venturebeat.com/2021/10/12/deepmind-is-developing-one-algorithm-to-rule-them-all/, 12 October 2021.
49. "From System 1 Deep Learning to System 2 Deep Learning", a talk delivered at the *Conference on Neural Information Processing Systems (NeurIPS 2019)* by Yoshua Bengio, published on https://youtu.be/T3sxeTgT4qc, slide 34.
50. "Mark Zuckerberg thinks Elon Musk is wrong on AI" by Sam Mattera, published by *The Motley Fool* on https://www.sfgate.com/business/fool/article/Mark-Zuckerberg-Thinks-Elon-Musk-is-Wrong-on-AI-7384131.php, 29 April 2016.

51. "9 failed scientific predictions", published on https://www.technologynetworks.com/tn/lists/10-failed-scientific-predictions-276945, 14 December 2016.
52. "The singularity: a philosophical analysis" by David Chalmers, 2010, contained in *The Singularity: Could Artificial Intelligence Really Out-Think Us (And Would We Want It To)?*, edited by Uziel Awret, published by Imprint Academic, 2016, p. 33.

Chapter 2. What is Intelligence?

1. "Biology vs physics: two ways of doing science" by Massimo Pigliucci, published in *The Philosophers' Magazine*, 21 December 2015.
2. *The Unity of Intelligence* by Frank Wilczek, published in *Possible Minds: Twenty-Five Ways of Looking at AI*, edited by John Brockman, published by Penguin Press, 2019, p. 66, citing *The Astonishing Hypothesis: The Scientific Search for the Soul* by Francis Crick, published by Touchstone, 1995.
3. *The Unity of Intelligence* by Frank Wilczek, published in *Possible Minds: Twenty-Five Ways of Looking at AI*, edited by John Brockman, published by Penguin Press, 2019, p. 64.
4. *Life 3.0: Being Human in the Age of Artificial Intelligence* by Max Tegmark, published by Allen Lane, 2017, p. 25.
5. *The Promise of Artificial Intelligence: Reckoning and Judgement* by Brian Cantwell Smith, published by MIT Press, 2019, p. 110.
6. Taken from an interview with Stuart Russell, published in Ford, op. cit., p. 63.
7. Bostrom, op. cit., p. 3.
8. *How the Mind Works* by Steven Pinker, published by Penguin Books in paperback, 1999, p. 62.
9. "Universal intelligence: a definition of machine intelligence" by Shane Legg and Marcus Hutter in *Minds & Machines*, vol. 17, 2007, p. 402.
10. "The Cattell-Horn-Carroll Model of Intelligence" by W. Joel Schneider and Kevin S. McGrew, published as chapter 4 of *Contemporary Intellectual Assessment: Theories, Tests, and Issues*, 3rd edition, edited by Dawn P. Flanagan and Patti L. Harrison, published by Guilford Press, 2012, p. 109.
11. Ibid., p. 100.

12. Legg and Hutter, op. cit., p. 399.
13. W.J. Schneider and K.S. McGrew, op. cit., p. 103.
14. W.J. Schneider and K.S. McGrew, op. cit., p. 105.
15. *The Measure of All Minds: Evaluating Natural and Artificial Intelligence* by José Hernández-Orallo, published by Cambridge University Press, 2017, pp. 67–68, citing "Age differences in fluid and crystallized intelligence" by J.L. Horn and R.B. Cattell, published in *Acta Psychologica*, vol. 26, pp. 107–129, 1967.
16. *Human Compatible: Artificial Intelligence and the Problem of Control* by Stuart Russell, published by Allen Lane, 2019, p. 34.
17. Bostrom, op. cit., p. 29.
18. *The Correspondence of Charles Darwin: Volume 2, 1837–1843* by Charles Darwin, edited by Frederick Burkhardt and Sydney Smith, published by Cambridge University Press, 1987.
19. *The Feeling of What Happens: Body, Emotion and the Making of Consciousness* by Antonio Damasio, published in paperback by Vintage, 2000, pp. 41–42.
20. *Predicting Human Decision-Making: From Prediction to Action* by Ariel Rosenfeld and Sarit Kraus, published by Morgan & Claypool, 2018, pp. 1–2.
21. *Discourse on the Method of Rightly Conducting One's Reason and of Seeking Truth in the Sciences* by Rene Descartes, 1637.
22. "What is it like to be a bat?" by Thomas Nagel, published in *The Philosophical Review*, vol. 83, no. 4, October 1974, p. 436.
23. *The Character of Consciousness* by David J. Chalmers, published by Oxford University Press, 2010, p. 8.

Chapter 3. Why We Need Human-Purposed AI

1. "High accuracy protein structure prediction using deep learning" by John Jumper, Richard Evans, Alexander Pritzel, Tim Green, Michael Figurnov, Kathryn Tunyasuvunakool, Olaf Ronneberger, Russ Bates, Augustin Žídek, Alex Bridgland, Clemens Meyer, Simon A A Kohl, Anna Potapenko, Andrew J Ballard, Andrew Cowie, Bernardino Romera-Paredes, Stanislav Nikolov, Rishub Jain, Jonas Adler, Trevor Back, Stig Petersen, David Reiman, Martin Steinegger, Michalina Pacholska, David Silver, Oriol Vinyals, Andrew W Senior, Koray Kavukcuoglu, Pushmeet Kohli, Demis Hassabis, published in *Fourteenth Critical Assessment of*

Techniques for Protein Structure Prediction (Abstract Book), 30 November–4 December 2020, pp. 22–24.

2. "Protein revolution" by Michael Le Page, published in *New Scientist*, 5 December 2020.
3. "Alibaba offered clients facial recognition to identify Uighur people, report reveals" by Helen Davidson, published on https://www.theguardian.com/business/2020/dec/17/alibaba-offered-clients-facial-recognition-to-identify-uighur-people, 17 December 2020.
4. *Army of None: Autonomous Weapons and the Future of War* by Paul Scharre, published by W.W. Norton & Company, 2018, pp. 46–48.
5. "The new killer robots" by Matthew Campbell, published in *The Sunday Times Magazine*, 14 November 2021.
6. Bostrom, op. cit., p. 23.
7. Bostrom, op. cit., p. 29.
8. Bostrom, op. cit., p. 22.
9. Russell, op. cit., p. 132.
10. "Artificial general intelligence and the human mental model" by Roman V. Yampolskiy and Joshua Fox, published in *Singularity Hypotheses: A Scientific and Philosophical Assessment*, edited by Amnon Eden, Johnny Søraker, James H. Moor and Eric Steinhart, published by Springer, 2012, pp. 1–2.
11. *2001: A Space Odyssey*, a film written by Stanley Kubrick and Arthur C. Clarke, directed and produced by Stanley Kubrick, 1968.
12. *Other Minds: The Octopus and the Evolution of Intelligent Life* by Peter Godfrey-Smith, published by William Collins, 2017, p. 67.
13. Yampolskiy and Fox, op. cit., p. 5.
14. "When will computer hardware match the human brain?" by Hans P. Moravec, published in *Journal of Evolution and Technology*, vol. 1, 1998.
15. "From mostly harmless to civilization-threatening: pathways to dangerous artificial intelligences" by Kaj Sotala, presented at *ECAP10: VIII European Conference on Computing and Philosophy*, edited by Klaus Mainzer; conference arranged by the Technical University of München, 2010.
16. "Deal or no deal? Training AI bots to negotiate" by Mike Lewis, Denis Yarats, Yann N Dauphin, Devi Parikh, Dhruv Batra (MAIR

researchers), on https://engineering.fb.com/2017/06/14/ml-applications/deal-or-no-deal-training-ai-bots-to-negotiate/, 14 June 2017.

17. "Masters of the universe", published in *The Economist*, 5 October 2019.
18. "Machine learning poses significant risks that need to be managed" by Mark Yallop, published in the *Financial Times*, 24 December 2019.
19. Yampolskiy and Fox, op. cit., abstract.
20. *The Precipice: Existential Risk and the Future of Humanity* by Toby Ord, published by Bloomsbury Publishing, 2020, p. 151.
21. Taken from an interview with Stuart Russell, published in Ford, op. cit., p. 63.
22. Bostrom, op. cit., p. 105.
23. Bostrom, op. cit., pp. 120-122.
24. "Learning from Tay's introduction" by Peter Lee, published on https://blogs.microsoft.com/blog/2016/03/25/learning-tays-introduction/, 25 March 2016.
25. *Humans*, created by Sam Vincent and Jonathan Brackley and produced by Channel 4, AMC Studios and Kudos for Channel 4 and AMC, 2015.
26. https://www.un.org/sustainabledevelopment/sustainable-development-goals/.
27. Bostrom, op. cit., pp. 145–148.
28. Bostrom, op. cit., pp. 129–131.
29. Scharre, op. cit., p. 351.
30. *AI Superpowers: China, Silicon Valley, and the New World War* by Kai-Fu Lee, published by Houghton Mifflin Harcourt Publishing Company, 2018.
31. Ibid., p. 83.
32. "Hot tropics" in *The Economist*, 4 December 2021.
33. "Son lays down gauntlet to Valley with Yahoo Japan's Line merger" by Kana Inagaki, published in the *Financial Times*, 19 November 2019.
34. Recommendation of the Council on Artificial Intelligence, OECD/LEGAL/0449, OECD, 2019.
35. "Artificial intelligence as a positive and negative factor in global risk" by Eliezer Yudkowsky, published in *Global Catastrophic Risks*,

edited by Nick Bostrom and Milan M. Ćirković, published by Oxford University Press, 2008.

36. Ibid., p. 11.
37. Ibid., p. 12.
38. "We wouldn't be able to control superintelligent machines" by Max-Planck-Gesellschaft, published on https://www.mpg.de/16231640/0108-bild-computer-scientists-we-wouldn-t-be-able-to-control-superintelligent-machines-149835-x, 11 January 2021.
39. Yudkowsky, op. cit., pp. 41–43.
40. "Whoever leads in AI will rule the world: Putin to Russian children on Knowledge Day", published on www.rt.com, 1 September 2017, https://www.rt.com/news/401731-ai-rule-world-putin/.
41. "China faces territorial issues with 18 nations; check details" by Zee Media Bureau, edited by Arun Kumar Chaubey, published on https://zeenews.india.com/world/china-faces-territorial-issues-with-18-nations-check-details-2292826.html, 30 June 2020.
42. "Codified crackdown" in *The Economist*, 11 September 2021.
43. *The Long Game: China's Grand Strategy to Displace American Order* by Rush Doshi, published by Oxford University Press, 2021, p. 289.
44. "The quantum computing threat to American security" by Arthur Herman, published in the *Wall Street Journal*, 10 November 2019.
45. *Chinese Public AI R&D Spending: Provisional Findings* by Ashwin Acharya and Zachary Arnold, published by The Center for Security and Emerging Technology, part of Georgetown's Walsh School of Foreign Service, 2019.
46. Taken from an interview with Peter Railton by Lucas Perry on moral learning and metaethics in AI systems, published by *Future of Life Institute* on https://futureoflife.org/2020/08/18/peter-railton-on-moral-learning-and-metaethics-in-ai-systems/, 18 August 2020.

Chapter 5. The Evolution of Autonomy: Control and Coordination

1. *The Deep History of Ourselves: The Four-Billion-Year Story of How We Got Conscious Brains* by Joseph LeDoux, published by Viking, 2019, p. 11.

2. "The senses of sea anemones: responses of the SS1 nerve net to chemical and mechanical stimuli" by I.D. McFarlane and I.D. Lawn, published in *Hydrobiologia*, vol. 216, 1991, pp. 599–604.
3. Godfrey-Smith, op. cit., p. 67.
4. "The secret of you" by Caroline Williams, published in *New Scientist*, 7 July 2018.
5. *An Introduction to Brain and Behaviour* by Bryan Kolb and Ian Q. Whishaw, fourth edition, published by Worth Publishers, 2014, p. 76.
6. "Equal numbers of neuronal and nonneuronal cells make the human brain an isometrically scaled-up primate brain" by Azevedo, F.A.C., Carvalho, L.R.B., Grinberg, L.T., Farfel, J.M., Ferretti, R.E.L., Leite, R.E.P., Filho, W.J., Lent, R. and Herculano-Houzel, S. in *Journal of Comparative Neurology*, vol. 513, issue 5, 10 April 2009, pp. 532–541.
7. "The elephant brain in numbers" by Suzana Herculano-Houzel, Kamilla Avelino-de-Souza, Kleber Neves, Jairo Porfírio, Débora Messeder, Larissa Mattos Feijó, José Maldonado and Paul R. Manger, published in *Frontiers in Neuroanatomy*, 12 June 2014.
8. *Cerebral Cortex: Principles of Operation* by Edmund T. Rolls, published in paperback by Oxford University Press, 2017, p. 227.
9. Ibid., Introduction, p. 32.
10. *The Astonishing Hypothesis: The Scientific Search for the Soul* by Francis Crick, published by Touchstone, 1995, pp. 98–103.
11. Kolb and Whishaw, op. cit., p. 141.
12. *The Organization of Behavior: A Neuropsychological Theory* by D.O. Hebb, published by Wiley & Sons, 1949.
13. "Separate visual pathways for perception and action" by Melvyn A. Goodale and A. David Milner, published in *Trends in Neurosciences*, vol. 15, no. 1, 1992.
14. "Hidden computational power found in the arms of neurons: the dendritic arms of some human neurons can perform logic operations that once seemed to require whole neural networks" by Jordana Cepelewicz, published on https://www.quantamagazine.org/neural-dendrites-reveal-their-computational-power-20200114/, 14 January 2020.
15. "Visual trick has AI mistake turtle for a gun", published in *New Scientist*, 11 November 2017, p. 19.

16. Rolls, op. cit., p. 17.

Chapter 6. Emotions

1. *Self Comes to Mind: Constructing the Conscious Brain* by Antonio Damasio, published by William Heinemann, 2010, p. 109.
2. *The Emotional Brain: The Mysterious Underpinnings of Emotional Life* by Joseph LeDoux, published in paperback by Phoenix, 1999, p. 40.
3. *Self Comes to Mind: Constructing the Conscious Brain* by Antonio Damasio, published by William Heinemann, 2010, p. 109.
4. *The Archaeology of Mind: Neuroevolutionary Origins of Human Emotions* by Jaak Panksepp and Lucy Biven, published by W.W. Norton & Company, Inc., 2012, p. 43.
5. "Both of us disgusted in *my* insula: the common neural basis of seeing and feeling disgust" by Bruno Wicker, Christian Keysers, Jane Plailly, Jean-Pierre Royet, Vittorio Gallese and Giacomo Rizzolatti, published in *Neuron*, vol. 40, pp. 655–664, 30 October 2003.
6. "Our restless minds" by Simon Baron-Cohen, published in *New Scientist*, 5 December 2020.
7. *Self Comes to Mind: Constructing the Conscious Brain* by Antonio Damasio, published by William Heinemann, 2010, pp. 122–123.
8. "Expression and the nature of emotion" by Paul Ekman, published as chapter 15 in *Approaches to Emotion*, edited by K. Scherer and P. Ekman, published by Psychology Press, 1984.
9. "Expression of doubt" by Douglas Heaven, published in *Nature*, vol. 578, 27 February 2020.
10. "Behind the smile" by Emma Young, published in *New Scientist*, 15 February 2020.
11. Panksepp and Biven, op. cit., p. xi.
12. *How We Learn: The New Science of Education and the Brain* by Stanislas Dehaene, published by Allen Lane, 2020, pp. 186–189.
13. Panksepp and Biven, op. cit., pp. 16–19.
14. Damasio, op. cit, pp. 125–126.
15. "Pride, personality, and the evolutionary foundations of human social status" by Joey T. Cheng, Jessica L. Tracy and Joseph Henrich, published in *Evolution and Human Behavior*, vol. 31, pp. 334–347, 2010, p. 336.

16. Panksepp and Biven, op. cit., p. 104.
17. "From experienced utility to decision utility" by Kent C. Berridge and John P. O'Doherty, published as chapter 18 in *Neuroeconomics: Decision-Making and the Brain*, 2nd edition, edited by Paul W. Glimcher and Ernst Fehr, published by Academic Press, an imprint of Elsevier Inc., 2014, p. 337.
18. "Dissecting components of reward: 'liking', 'wanting', and learning" by Kent C. Berridge, Terry E. Robinson and J. Wayne Aldridge, published in *Current Opinion in Pharmacology*, vol. 9, 2009, pp. 65–73.
19. "Temporally restricted dopaminergic control of reward-conditioned movements" by Kwang Lee, Leslie D. Claar, Ayaka Hachisuka, Konstantin I. Bakhurin, Jacquelyn Nguyen, Jeremy M. Trott, Jay L. Gill and Sotiris C. Masmanidis, published in *Nature Neuroscience*, vol. 23, February 2020, pp. 209–216.
20. Berridge, Robinson and Aldridge, op. cit., p. 65.
21. *The Emotional Brain: The Mysterious Underpinnings of Emotional Life* by Joseph LeDoux, published in paperback by Phoenix, 1999, pp. 157–178.
22. Panksepp and Biven, op. cit., p. 182.
23. "Neural correlates of admiration and compassion" by M.H. Immordino-Yang, A. McColl, H. Damasio and A. Damasio, published in *Proceedings of the National Academy of Sciences*, vol. 106, no. 19, 2009, pp. 8021–8026.
24. "Brain basis of human social interaction: from concepts to brain imaging" by Riitta Hari and Miiamaaria V. Kujala, published in *Physiological Reviews*, vol. 89, no. 2, 1 April 2009, pp. 465–466.
25. "Regret and its avoidance: a neuroimaging study of choice behavior" by G. Coricelli, H.D. Critchley, M. Joffily, J.P. O'Doherty, A. Sirigu and R.J. Dolan, published in *Nature Neuroscience*, vol. 8, 7 August 2005, pp. 1255–1262.
26. "When your gain is my pain and your pain is my gain: neural correlates of envy and schadenfreude" by H. Takahashi, M. Kato, M. Matsuura, D. Mobbs, T. Suhara and Y. Okubo, 2009, published in *Science*, vol. 323, pp. 937–939. Cited in Glimcher and Fehr, op. cit., p. 526.
27. "The green-eyed monster and malicious joy: the neuroanatomical bases of envy and gloating (schadenfreude)" by Simone G.

Shamay-Tsoory, Yasmin Tibi-Elhanany and Judith Aharon-Peretz, published in *Brain*, vol. 130, June 2007, pp. 1663–78. Cited in Glimcher and Fehr, op. cit., p. 526.

28. *The Righteous Mind: Why Good People are Divided by Politics and Religion* by Jonathan Haidt, published by Allen Lane, 2012, pp. 62–63, 332.
29. "The amygdala and ventromedial prefrontal cortex in morality and psychopathy" by R.J.R. Blair, published in *Trends in Cognitive Sciences*, vol. 11, issue 9, September 2007, pp. 387–392.
30. "A cognitive neuroscience perspective on psychopathy: evidence for paralimbic system dysfunction" by Kent A. Kiehl, published in *Psychiatry Research*, vol. 142, issues 2–3, 15 June 2006, pp. 107–128.
31. "Why do we care" by Patricia Churchland, published in *New Scientist*, 28 September 2019.
32. https://www.caressesrobot.org/en/project/.
33. "Robots get personal to boost mental health of care home residents" by Greg Hurst, published in *The Times*, 8 September 2020.

Chapter 7. Cognition

1. *The Bonobo and the Atheist: In Search of Humanism Among the Primates* by Frans de Waal, published in paperback by W.W. Norton & Co. Inc., 2014, p. 129.
2. *The Deep History of Ourselves: The Four-Billion-Year Story of How We Got Conscious Brains* by Joseph LeDoux, published by Viking, 2019, p. 34.
3. *The History of Philosophy* by A.C. Grayling, published by Viking, 2019, p. 219.
4. Ibid., p. 247.
5. Ibid., p. 197.
6. "Memory systems of the brain: a brief history and current perspective" by Larry R Squire, published in *Neurobiology of Learning and Memory*, vol. 82, issue 3, November 2004, pp. 171–177.
7. *Elements of Episodic Memory* by Endel Tulving, published by Oxford University Press, 1983.
8. Squire, op. cit., p. 174.
9. "The evolution of foresight: what is mental time travel, and is it unique to humans?" by Thomas Suddendorf and Michael C.

Corballis, published in *Behavioural and Brain Sciences*, vol. 30, issue 3, June 2007, pp. 299–313.

10. "Tracking the construction of episodic future thoughts" by Arnaud D'Argembeau and Arnaud Mathy, published in *Journal of Experimental Psychology: General*, vol. 140, no. 2, pp. 258–271, 2011.
11. Ibid., p. 268.
12. "Possible selves" by H. Markus and P. Nurius, published in *American Psychologist*, vol. 41, pp. 954–969, 1986.
13. *Cognition, Brain, and Consciousness: Introduction to Cognitive Neuroscience* by Bernard J. Baars and Nicole M. Gage, 2nd edition, published by Academic Press, 2010, p. 59.
14. Ibid., chap. 1, p. 8.
15. "The magical number 4 in short-term memory: a reconsideration of mental storage capacity" by Nelson Cowan, published in *Behavioral and Brain Sciences*, vol. 24 (1), 2000, pp. 87–114.
16. "Chimps outperform humans at memory task" by Rowan Hooper, published on https://www.newscientist.com/article/dn12993-chimps-outperform-humans-at-memory-task/, 3 December 2007.
17. *How We Learn: The New Science of Education and the Brain* by Stanislas Dehaene, published by Allen Lane, 2020, p. 53.
18. "Intuitive physics: current research and controversies" by James R. Kubricht, Keith J. Holyoak and Hongjing Lu, published in *Trends in Cognitive Sciences*, vol. 21, issue 10, pp. 749–759, 1 October 2017.
19. *Artificial Intelligence: A Guide for Thinking Humans* by Melanie Mitchell, published by Pelican Books, 2019, p. 211.
20. "Single-cell recognition: a Halle Berry brain cell" by Marcus Woo, published on https://www.caltech.edu/about/news/single-cell-recognition-halle-berry-brain-cell-1013, 16 June 2005.
21. "Brain cells for grandmother" by Rodrigo Quian Quiroga, Itzhak Fried and Christof Koch, *Scientific American*, published on www.ScientificAmerican.com, February 2013.
22. "The neurobiological foundation of memory retrieval" by Paul W. Frankland, Sheena A. Josselyn and Stefan Köhler, published in *Nature Neuroscience*, vol. 22, October 2019, pp. 1576–1585.
23. Taken from an interview with Gary Marcus, published in Ford, op. cit., pp. 307–308.

24. "One of these is a power drill" by Chris Baraniuk, published in *New Scientist*, 27 April 2019.
25. *Thinking, Fast and Slow* by Daniel Kahneman, published in paperback by Penguin Books, 2012, pp. 20–21.
26. Ibid., p. 52.
27. Ford, op. cit., p. 78.
28. *Die Another Day*, produced by Michael G. Wilson and Barbara Broccoli with Eon Productions and Metro-Goldwyn-Mayer Pictures, directed by Lee Tamahori, 2002.
29. Kahneman, op. cit., p. 20.
30. *Rationality & the Reflective Mind* by Keith E. Stanovich, published by Oxford University Press, 2011, pp. 142–144.
31. *The Book of Why: The New Science of Cause and Effect* by Judea Pearl and Dana Mackenzie, published in paperback by Penguin Books, 2019, pp. 167–188.
32. Stanovich, op. cit., p. 69.
33. "How language helps us think" by Ray Jackendoff, published in *Pragmatics and Cognition*, vol. 4, issue 1, 1996, p. 10.
34. "Ravens parallel great apes in flexible planning for tool-use and bartering" by C. Kabadayi and M. Osvath, published in *Science*, 14 July 2017, issue 6347, pp. 202–204.
35. *The Language Instinct: The New Science of Language and Mind* by Steven Pinker, published by The Penguin Group, 1994.
36. Ibid., pp. 78–82.
37. *Why Only Us: Language and Evolution* by Robert C. Berwick and Noam Chomsky, published by MIT Press, 2016, p. 50.
38. "Concepts in a probabilistic language of thought" by Noah D. Goodman, Joshua B. Tenenbaum, Tobias Gerstenberg, published as chapter 22 of *The Conceptual Mind: New Directions in the Study of Concepts* edited by Eric Margolis and Stephen Laurence, published by Massachusetts Institute of Technology, 2015.
39. *Lost in Space*, created and produced by Irwin Allen for CBS, 1965, as published on https://en.wikipedia.org/wiki/Lost_in_Space.
40. Domingos, op. cit., p. 61.
41. "Task representations in neural networks trained to perform many cognitive tasks" by Guangyu Robert Yang, Madhura R. Joglekar, H. Francis Song, William T. Newsome and Xiao-Jing

Wang, published in *Nature Neuroscience*, vol. 22, February 2019, pp. 297–306.

42. *The Mind Doesn't Work That Way: The Scope and Limits of Computational Psychology* by Jerry Fodor, published in paperback by MIT Press, 2001, p. 41.
43. Roitblat, op. cit., pp. 36–37.
44. Fodor, op. cit., chap. 2, p. 38.
45. Ford, op. cit., pp. 125–126.
46. Taken from interviews with Yoshua Bengio, Geoffrey Hinton and Yann LeCun, published in Ford, op. cit., pp. 22, 84–85, 127, respectively.
47. Taken from an interview with Yann LeCun, published in Ford, op. cit., p. 127.
48. "From System 1 Deep Learning to System 2 Deep Learning", a talk delivered at the *Conference on Neural Information Processing Systems (NeurIPS 2019)* with Yoshua Bengio, https://youtu.be/T3sxeTgT4qc. Also taken from an interview with Gary Marcus, published in Ford, op. cit., p. 22.
49. Roitblat, op. cit., p. 227.
50. "Neuro-symbolic A.I. is the future of artificial intelligence. Here's how it works" by Luke Dormehl, published on https://www.digitaltrends.com/cool-tech/neuro-symbolic-ai-the-future/, 5 January 2020.
51. "This know-it-all AI learns by reading the entire web nonstop" by Will Douglas Heaven, published on https://www.technologyreview.com/2020/09/04/1008156/knowledge-graph-ai-reads-web-machine-learning-natural-language-processing/, 4 September 2020.
52. "Sounds of action: using ears, not just eyes, improves robot perception" by Byron Spice, published on https://www.cs.cmu.edu/news/2020/sounds-action-using-ears-not-just-eyes-improves-robot-perception, 14 August 2020.
53. "Competitive programming with AlphaCode" by the AlphaCode Team, 2 February 2022, published on https://deepmind.com/blog/article/Competitive-programming-with-AlphaCode.
54. https://codeforces.com/.

Chapter 8. Decision-Making

1. "Value-based decision-making" by Paul W. Glimcher, published as chapter 20 in *Neuroeconomics: Decision-Making and the Brain*, 2nd edition, edited by Paul W. Glimcher and Ernst Fehr, published by Academic Press, an imprint of Elsevier Inc., 2014.
2. "Multiple systems for value learning" by Nathaniel D. Daw and John P. O'Doherty, published in Glimcher and Fehr, op. cit., chapter 21.
3. "Actions and habits: the development of a behavioural autonomy" by A. Dickinson, published in *Philosophical Transactions of the Royal Society B Biological Sciences*, vol. 308, issue 1135, pp. 67–78, 1985.
4. "Bidirectional instrumental conditioning" by A. Dickinson, published in *Quarterly Journal of Experimental Psychology Section B*, vol. 49, issue 4b, pp. 289–306, 1996.
5. "Variations in the sensitivity of instrumental responding to reinforcer devaluation" by C. Adams, published in *Quarterly Journal of Experimental Psychology Section B*, vol. 34, issue 2b, pp. 77–98, 1982.
6. "Goal-directed instrumental action: contingency and incentive learning and their cortical substrates" by B.W. Balleine and A. Dickinson, published in *Neuropharmacology*, vol. 37, issues 4–5, 5 April 1998, pp. 407–419.
7. Daw and O'Doherty, op. cit., pp. 396–397.
8. "A specific role for posterior dorsolateral striatum in human habit learning" by E. Tricomi, B.W. Balleine and J.P. O'Doherty, published in *European Journal of Neuroscience*, vol. 29, issue 11, pp. 2225–2232, 2009.
9. "Valuation, intertemporal choice, and self-control" by Joseph W. Kable, published in Glimcher and Fehr, op. cit., chapter 10.
10. "Value-based decision-making" by Paul W. Glimcher, published in Glimcher and Fehr, op. cit., chapter 20, pp. 386–388.
11. "Multistage valuation signals and common neural currencies" by Michael L. Platt and Hilke Plassmann, published in Glimcher and Fehr, op. cit., chapter 13, pp. 242–244.
12. Glimcher, op. cit., in Glimcher and Fehr, op. cit., pp. 386–388.
13. "Integrating benefits and costs in decision-making" by Jonathan D. Wallis and Matthew F.S. Rushworth, published in Glimcher and Fehr, op. cit., chapter 22, pp. 419–421.

14. "Teaching the cerebellum about reward" by Javier F. Medina, published in *Nature Neuroscience*, vol. 22, June 2019, pp. 846–848.
15. "Coordinated cerebellar climbing fiber activity signals learned sensorimotor predictions" by William Heffley, Eun Young Song, Ziye Xu, Benjamin N. Taylor, Mary Anne Hughes, Andrew McKinney, Mati Joshua and Court Hull, published in *Nature Neuroscience*, vol. 21, October 2018, pp. 1431–1441.
16. "Predictive and reactive reward signals conveyed by climbing fiber inputs to cerebellar Purkinje cells" by Dimitar Kostadinov, Maxime Beau, Marta Blanco Pozo and Michael Häusser, published in *Nature Neuroscience*, vol. 22, June 2019, pp. 950–962.
17. "Neuroeconomics of emotion and decision making" by Karolina M. Lempert and Elizabeth A. Phelps, published in Glimcher and Fehr, op. cit., chapter 12, p. 228.
18. *Concise Oxford English Dictionary*, eleventh edition (revised), published by Oxford University Press, p. 1419, 2006.
19. Glimcher, op. cit., in Glimcher and Fehr, op. cit., pp. 377–384.
20. Rolls, op. cit., chapter 5, p. 135.
21. Glimcher, op. cit., in Glimcher and Fehr, op. cit., p. 386.
22. "Deep and beautiful. The reward prediction error hypothesis of dopamine" by Matteo Colombo, published in *Studies in History and Philosophy of Science Part C: Studies in History and Philosophy of Biological and Biomedical Sciences*, vol. 45, March 2014, pp. 57–67.
23. Mitchell, op. cit., pp. 162–187.
24. "Advanced reinforcement learning" by Nathaniel D. Daw, published in Glimcher and Fehr, op. cit., chapter 16, p. 316.
25. "Multiple systems for value learning" by Nathaniel D. Daw and John P. O'Doherty, published in Glimcher and Fehr, op. cit., chapter 21., pp. 398–400.
26. Ibid., p. 395.
27. "Building machines that learn and think for themselves: commentary on Lake, Ullman, Tenenbaum, and Gershman, Behavioral and Brain Sciences, 2017" by M. Botvinick, D.G.T. Barrett, P. Battaglia, N. de Freitas, D. Kumaran, J.Z. Leibo, T. Lillicrap, J. Modayil, S. Mohamed, N.C. Rabinowitz, D.J. Rezende, A. Santoro, T. Schaul, C. Summerfield, G. Wayne, T. Weber, D. Wierstra, S. Legg, and D. Hassabis, published by *DeepMind* on https://arxiv.org/ftp/arxiv/papers/1711/1711.08378.pdf, 22 November 2017.

28. Kahneman, op. cit.
29. Taken from an interview with Yann LeCun, published in Ford, op. cit., p. 126.
30. Daw and O'Doherty, op. cit., in Glimcher and Fehr, op. cit., p. 405.
31. *Concise Oxford English Dictionary*, eleventh edition (revised) 2006, published by Oxford University Press.
32. Rolls, op. cit., p. 440.
33. Rolls, op. cit., p. 129.
34. Rolls, op. cit., p. 131.
35. *The Better Angels of Our Nature* by Steven Pinker, published by the Penguin Group, 2011.

Chapter 10. Language, Thought and Consciousness

1. *Homo Deus: A Brief History of Tomorrow* by Yuval Noah Harari, published in paperback by Vintage, 2017, p. 332.
2. "Chasing the rainbow: the non-conscious nature of being" by D.A. Oakley and P.W. Halligan, published in *Frontiers in Psychology*, vol. 8, article 1924, 14 November 2017, p. 7.
3. *The Feeling of Life Itself: Why Consciousness is Widespread but Can't Be Computed* by Christof Koch, published by MIT Press, 2019, p. 30.
4. "How language helps us think" by Ray Jackendoff, published in *Pragmatics and Cognition*, 1996, p. 2.
5. *The Language Instinct: The New Science of Language and Mind* by Steven Pinker, published in paperback by The Penguin Group, 1995, p. 82.
6. *A User's Guide to Thought and Meaning* by Ray Jackendoff, published by Oxford University Press, 2012, p. 85.
7. *Language: The Cultural Tool* by Daniel Everett, published in paperback by Profile Books Ltd., 2012, pp. 168–169.
8. *How Language Began: The Story of Humanity's Greatest Invention* by Daniel Everett, published by Profile Books Ltd., 2017, pp. 50–64.
9. *Animal Languages: The Secret Conversations of the Living World* by Eva Meijer, published by John Murray (Publishers), 2019, p. 27.
10. *The Origin of Consciousness in the Breakdown of the Bicameral Mind* by Julian Jaynes, published in paperback by Houghton Mifflin Company, 1982.

11. *Wider than the Sky: A Revolutionary View of Consciousness* by Gerald Edelman, published by Yale University Press, 2004.
12. "Towards a conversational agent that can chat about...anything", posted by Daniel Adiwardana and Thang Luong, Google Research Brain Team on https://ai.googleblog.com/2020/01/towards-conversational-agent-that-can.html, 28 January 2020.
13. Transcript extracted from *Real Talk: Michael Rosen Talks to Conversation Analyst Elizabeth Stokoe about the Science of Talk on Word of Mouth*, released on BBC Radio 4, 20 February 2020.
14. "COMET: commonsense transformers for automatic knowledge graph construction" by Antoine Bosselut, Hannah Rashkin, Maarten Sap, Chaitanya Malaviya, Asli Celikyilmaz, Yejin Choi, published on https://arxiv.org/abs/1906.05317, 14 June 2019.
15. "Common sense comes closer to computers" by John Pavlus, published in *Quanta Magazine*, 30 April 2020.

Chapter 11. Selected Theories of Consciousness

1. *The Character of Consciousness* by David J. Chalmers, published by Oxford University Press, 2010, p. 5.
2. "What is it like to be a bat?" by Thomas Nagel, published in *The Philosophical Review*, vol. 83, no. 4, October 1974.
3. Chalmers, op. cit., p. 4.
4. Created by Malcolm Judge in 1962, subsequently drawn by John Dallas from 1989, published by DC Thomson.
5. "Symmetry of snowflakes" by Ian Stewart, published on https://warwick.ac.uk/newsandevents/knowledge-archive/science/snowflakes/, September 2010.
6. "Strong and weak emergence" by David Chalmers, being chapter 11 of *The Re-emergence of Emergence: The Emergentist Hypothesis from Science to Religion*, edited by Philip Clayton and Paul Davies, published by Oxford University Press, 2006.
7. *Language, Consciousness, Culture: Essays on Mental Structure* by Ray Jackendoff, published by MIT Press, 2007, pp. 83–85.
8. *Freedom & Neurobiology: Reflections on Free Will, Language, and Political Power* by John R. Searle, published by Columbia University Press, 2007, pp. 69–70.
9. *Becoming Artificial: A Philosophical Exploration into Artificial Intelligence and What It Means to Be Human* by Danial Sonik and

Alessandro Colarossi, published by Imprint Academic, 2020, pp. 43–47.
10. *The Character of Consciousness* by David J. Chalmers, published by Oxford University Press, 2010, p. 137.
11. Chalmers, op. cit., p. 138.
12. *Consciousness Explained* by Daniel Dennett, published in paperback by Penguin Books, 1993.
13. Ibid., pp. 33–39.
14. Ibid., pp. 111–115.
15. Ibid., pp. 193–199.
16. Ibid., pp. 225–226.
17. Ibid., pp. 219–220.
18. Ibid., pp. 222–226 and pp. 277–280.
19. Chalmers, op. cit., pp. 105–111.
20. Dennett, op. cit., pp. 219–220.
21. *Mind Time: The Temporal Factor in Consciousness* by Benjamin Libet, published by Harvard University Press, 2004, pp. 123–124.
22. Ibid., pp. 137–149.
23. Ibid., p. 108.
24. Ibid., p. 147.
25. Ibid., p. 159.
26. "Higher-order theories of consciousness: an overview" by Rocco J. Gennaro, from *Higher-Order Theories of Consciousness: An Anthology*, edited by Rocco J. Gennaro, published by John Benjamins Publishing Company, 2004.
27. *Cerebral Cortex: Principles of Operation* by Edmund T. Rolls, published by Oxford University Press, in paperback, 2017, pp. 129–130.
28. Ibid., p. 131.
29. Ibid., p. 131 and pp. 425–426.
30. "Two concepts of consciousness" by D. Rosenthal, published in *Philosophical Studies*, May 1986, pp. 329–359.
31. Rolls, op. cit., p. 426.
32. Ibid., pp. 423–426.
33. Ibid., p. 424.
34. *A Cognitive Theory of Consciousness* by Bernard J. Baars, published by Cambridge University Press, 1989.

35. "A neuronal model of a global workspace in effortful cognitive tasks" by Stanislas Dehaene, Michel Kerszberg and Jean-Pierre Changeux, published by *Proceedings of the National Academy of Sciences*, vol. 95, pp. 14529–14534, November 1998.
36. *Cognition, Brain, and Consciousness: Introduction to Cognitive Neuroscience* by Bernard J. Baars and Nicole M. Gage, 2nd edition, published by Academic Press, 2010, p. 287.
37. *Consciousness and the Brain: Deciphering How the Brain Codes Our Thoughts* by Stanislas Dehaene, published by Penguin Books, 2014, p. 163.
38. Ibid., p. 171.
39. Baars and Gage, op. cit., p. 288.
40. *Consciousness: A Very Short Introduction* by Susan Blackmore, published by Oxford University Press, 2017, p. 48.
41. Dehaene, op. cit., p. 168.
42. Dehaene, op. cit., pp. 166–168.
43. Dehaene, op. cit., p. 165.
44. Searle, op. cit., p. 26.
45. https://en.wikipedia.org/wiki/Turtles_all_the_way_down.
46. *Galileo's Error: Foundations for a New Science of Consciousness* by Philip Goff, published by Rider, an imprint of Ebury Publishing, 2019, p. 115.
47. Ibid., pp. 135–137.
48. From Boris Johnson, https://www.brainyquote.com/quotes/boris_johnson_452516.
49. Goff, op. cit., p. 126.
50. Goff, op. cit., p. 180.
51. Goff, op. cit., p. 144.
52. Goff, op. cit., p. 164.
53. "Consciousness and the laws of physics" by Sean Carroll, published in *Journal of Consciousness Studies*, vol. 28, no. 9–10, 2021, p. 29.
54. *Self Comes to Mind: Constructing the Conscious Brain* by Antonio Damasio, published by William Heinemann, 2010, p. 190.
55. Ibid., p. 257.
56. Ibid., p. 202.
57. Ibid., pp. 201–205.
58. Ibid., p. 201.

59. Ibid., p. 203.
60. Ibid., p. 9.
61. Ibid., p. 18.
62. *The Feeling of What Happens: Body, Emotion and the Making of Consciousness* by Antonio Damasio, published in paperback by Vintage, 2000, p. 176.
63. *Self Comes to Mind: Constructing the Conscious Brain* by Antonio Damasio, published by William Heinemann, 2010, pp. 210–240.
64. Ibid., p. 259.
65. Ibid., p. 268.
66. Ibid., p. 257.
67. "An information integration theory of consciousness" by Giulio Tononi, published in *BMC Neuroscience*, 2 November 2004.
68. *The Feeling of Life Itself: Why Consciousness is Widespread but Can't Be Computed* by Christof Koch, published by MIT Press, 2019, p. 74.
69. "Consciousness: here, there and everywhere?" by Giulio Tononi and Christof Koch, published in *Philosophical Translations Royal Society B Biological Sciences*, vol. 370, issue 1668, 2015, p. 7.
70. "Integrated information theory: from consciousness to its physical substrate" by Giulio Tononi, Melanie Boly, Marcello Massimini and Christof Koch, published in *Nature Reviews Neuroscience*, vol. 17, 2016, p. 450.
71. Tononi and Koch, op. cit., p. 11.
72. Tononi and Koch, op. cit., p. 9.
73. Koch, op. cit., pp. 76–77.
74. Koch, op. cit., p. 159.
75. "The nature of consciousness experiments found to largely determine their results" by Tel Aviv University, published on https://medicalxpress.com/news/2022-03-nature-consciousness-largely-results.html, 16 March 2022.

Chapter 12. Other Considerations with Respect to Consciousness

1. "Can lab-grown brains become conscious" by Sara Reardon, published in *Nature*, 27 October 2020.
2. *Self Comes to Mind: Constructing the Conscious Brain* by Antonio Damasio, published by William Heinemann, 2010, p. 3.

3. Ibid., p. 163.
4. *The Character of Consciousness* by David J. Chalmers, published by Oxford University Press, 2010, pp. 20–23.
5. Ibid., p. 20.
6. Ibid., p. 22.
7. "The attentional requirements of consciousness" by Michael A. Cohen, Patrick Cavanagh, Marvin M. Chun and Ken Nakayama, published in *Trends in Cognitive Sciences*, August 2016, vol. 16, no. 8, p. 411.
8. *Consciousness and the Brain: Deciphering How the Brain Codes Our Thoughts* by Stanislas Dehaene, published by Penguin Books, 2014, p. 77.
9. *Cognition, Brain, and Consciousness: Introduction to Cognitive Neuroscience* by Bernard J. Baars and Nicole M. Gage, 2nd edition, published by Academic Press, 2010, p. 274.
10. Dehaene, op. cit., p. 75.
11. *The Tell-Tale Brain: Unlocking the Mystery of Human Nature* by V.S. Ramachandran, published by William Heinemann, 2011, p. 63.
12. Baars and Gage, op. cit., p. 287.
13. *The Illusion of Conscious Will* by Daniel M. Wegner, published by MIT Press, 2002, p. 221.
14. Wegner, op. cit., p. 269.
15. *Consciousness and the Social Brain* by Michael S.A. Graziano, published in paperback by Oxford University Press, 2015, p. 42.
16. Wegner, op. cit., p. 64.
17. "Apparent mental causation: sources of the experience of mental will" by D.M. Wegner and T. Wheatley, published in *American Psychologist*, 54, 1999, pp. 480–491.
18. Wegner, op. cit., p. 341.
19. "Guilt: an interpersonal approach" by R.F. Baumeister, A.M. Stillwell and T.F. Heatherton, published in *Psychological Bulletin*, vol. 115, March 1994, pp. 243–267.
20. "The moral emotions" by Jonathan Haidt, published in *Handbook of Affective Sciences*, edited by R.J. Davidson, K. Scherer and H.H. Goldsmith, published by Oxford University Press, 2001.
21. "Hangdog Fido plays on owner's feelings" by Kevin Dowling, published in *The Sunday Times*, 4 January 2015.

22. *The Primordial Emotions: The Dawning of Consciousness* by Derek Denton, published by Oxford University Press, 2005, p. 7.
23. Godfrey-Smith, op. cit., pp. 92–97.
24. Edelman, op. cit., p. 8.
25. Edelman, op. cit., p. 132.
26. Panksepp and Biven, op. cit., p. 13.
27. Ibid., p. 32.
28. Edelman, op. cit., pp. 78–86.
29. Edelman, op. cit., p. 85.
30. Dennett, op. cit., p. 378.
31. Grayling, op. cit., p. 220.
32. *Self Comes to Mind: Constructing the Conscious Brain* by Antonio Damasio, published by William Heinemann, 2010, pp. 167–168.
33. Ibid., pp. 168–169.

Chapter 14. Decision-Making and Complexity

1. *The Emotional Brain: The Mysterious Underpinnings of Emotional Life* by Joseph LeDoux, published in paperback by Phoenix, 1999, p. 143.
2. Ibid., p. 149.
3. "Multiple systems for value learning" by Nathaniel D. Daw and John P. O'Doherty, published as chapter 21 in *Neuroeconomics: Decision-Making and the Brain*, 2nd edition, edited by Paul W. Glimcher and Ernst Fehr, published by Academic Press, an imprint of Elsevier Inc., 2014, p. 395.
4. "The plant whisperer" by Joshua Howgego, published in *New Scientist*, 24 November 2018.
5. Daw and O'Doherty, op. cit., in Glimcher and Fehr, op. cit., p. 395.
6. "Actions and habits: the development of a behavioural autonomy" by Anthony Dickinson, published in *Philosophical Transactions of the Royal Society B Biological Sciences*, vol. 308, no. 1135, 13 February 1985, pp. 67–78.
7. Daw and O'Doherty, op. cit., in Glimcher and Fehr, op. cit., pp. 395–396.
8. Daw and O'Doherty, op. cit., in Glimcher and Fehr, op. cit., p. 396.
9. "Animal magic" by Sam Wong, published in *New Scientist*, 18 December 2021.

10. "Compound tool construction by New Caledonian crows" by A.M.P. von Bayern, S. Danel, A.M.I. Auersperg, B. Mioduszewska and A. Kacelnik, published in *Scientific Reports*, 8:15676, 24 October 2018.
11. Ibid., p. 4.
12. Daw and O'Doherty, op. cit., in Glimcher and Fehr, op. cit., pp. 398–400.
13. von Bayern *et al.*, op. cit., p. 5.
14. *Baboon Metaphysics: The Evolution of a Social Mind* by Dorothy L. Cheney and Robert M. Seyfarth, published by The University of Chicago Press, 2007.
15. Ibid., p. 116.
16. Ibid., p. 197.
17. *Brainstorming: Views and Interviews on the Mind* by Shaun Gallagher, published by Imprint Academic, 2008, pp. 184–185.
18. Cheney and Seyfarth, op. cit., p. 278.
19. Cheney and Seyfarth, op. cit., p. 279.
20. Cheney and Seyfarth, op. cit., p. 279.
21. "Field experiments find no evidence that chimpanzee nut cracking can be independently innovated" by Kathelijne Koops, Aly Gaspard Soumah, Kelly L. van Leeuwen, Henry Didier Camara and Tetsuro Matsuzawa, published in *Nature Human Behaviour*, 24 January 2022.
22. "Chimps crack the secret of evolution" by Tom Whipple, published in *The Times*, 25 January 2022.
23. A US legal term used in certain circumstances to place issues clearly and objectively in one category or another, from Wikipedia: https://en.wikipedia.org/wiki/Bright-line_rule.
24. *The Crimean War: A History* by Orlando Figes, published in paperback by Picador, 2012, pp. 247–248.
25. Ibid., p. 248.
26. Ibid., p. 252.
27. *The Reason Why* by Cecil Woodham Smith, published by Richard Clay and Company Ltd., p. 243.
28. *The Somme* by Richard van Emden, published by Pen & Sword Military, 2016, p. 226.
29. Ibid., pp. 242–243.
30. Ibid., p. 117.

Chapter 15. Self

1. *Self Comes to Mind: Constructing the Conscious Brain* by Antonio Damasio, published by William Heinemann, 2010, p. 9.
2. *The Ego Trick* by Julian Baggini, published by Granta Publications, 2011, p. 119.
3. "Philosophical conceptions of the self: implications for cognitive science" by Shaun Gallagher, published in *Trends in Cognitive Sciences*, vol. 4, no. 1, January 2000, p. 16.
4. *The Deep History of Ourselves: The Four-Billion-Year Story of How We Got Conscious Brains* by Joseph LeDoux, published by Viking, 2019, p. 303.
5. "How language helps us think" by Ray Jackendoff, published in *Pragmatics and Cognition*, 1996, p. 10.
6. "On agency and body-ownership: phenomenological and neuro-cognitive reflections" by Manos Tsakiris, Simone Schütz-Bosbach and Shaun Gallagher, published in *Consciousness and Cognition*, 2007, p. 2.
7. "Intrinsic architecture underlying the relations among the default, dorsal attention, and frontoparietal control networks of the human brain" by R. Nathan Spreng, Jorge Sepulcre, Gary R. Turner, W. Dale Stevens and Daniel L. Schacter, published in *Journal of Cognitive Neuroscience*, vol. 25(1), January 2013, pp. 74–86.
8. *Affect Regulation and the Origin of the Self: The Neurobiology of Emotional Development* by Allan N. Schore, published by Routledge in the Classic Edition, 2016, p. 10.
9. "Introduction to structure-function relationships in the developing brain" by J.L. Noebels, 1989, in *Problems and Concepts in Developmental Neurophysiology*, edited by P. Kellaway and J.L. Noebels, published by Johns Hopkins University Press, p. 152.
10. Schore, op. cit., pp. 24–25.
11. "Critical periods in psychoanalytic theories of personality development" by B. Beit-Hallahmi, published in *Sensitive Periods in Development: Interdisciplinary Perspectives*, edited by M.H. Bornstein, published by Lawrence Erlbaum Associates, pp. 211–221.
12. *Why Love Matters: How Affection Shapes a Baby's Brain* by Sue Gerhardt, 2nd edition, published by Routledge, 2015, p. 11.
13. Schore, op. cit., p. 71.

14. Schore, op. cit., pp. 71–72.
15. Schore, op. cit., p. 92.
16. Schore, op. cit., pp. 200–201.
17. "Moral internalization, parental power, and the nature of parent-child interaction" by M.L. Hoffman, published in *Developmental Psychology*, vol. 11, 1975, p. 236.
18. Schore, op. cit., p. 230.
19. Schore, op. cit., p. 238.
20. Schore, op. cit., p. 239.
21. *The Philosophical Baby: What Children's Minds Tell Us about Truth, Love & the Meaning of Life* by Alison Gopnik, published by The Bodley Head, 2009, p. 146.
22. Ibid., p. 59.
23. Ibid., p. 151.
24. Ibid., p. 147.
25. Ibid., p. 156.
26. "Lucy's species heralded the rise of long childhoods in hominids" by Bruce Bower, published in *Science News* on https://www.sciencenews.org/article/lucy-species-brain-skull-heralded-rise-long-childhoods-hominids, 1 April 2020.
27. Kahneman, op. cit., pp. 43–44.

Chapter 16. Mirror Neurons

1. "Understanding motor events: a neurophysiological study" by G. di Pellegrino, L. Fadiga, L. Fogassi, V. Gallese, and G. Rizzolatti, published in *Experimental Brain Research*, vol. 91, 14 July 1992, pp. 176–180.
2. *Mirrors in the Brain – How Our Minds Share Actions and Emotions* by Giacomo Rizzolatti and Corrado Sinigaglia, published by Oxford University Press, 2008, pp. 79–80.
3. Hari and Kujala, op. cit., p. 461.
4. Rizzolatti and Sinigaglia, op. cit., pp. 97–100.
5. Ibid., p. 144.
6. Ibid., pp. 190–191.
7. Ibid., pp. 170–171.
8. Ibid., p. 124.
9. Ibid., p. 100.
10. Ibid., p. 98.

11. *Mirroring People: The Science of Empathy and How We Connect with Others* by Marco Iacoboni, published in paperback by Picador, 2009, p. 12.
12. Rizzolatti and Sinigaglia, op. cit., p. 80.
13. Ibid., p. 84.
14. Ibid., p. 82.
15. Iacoboni, op. cit., p. 26.
16. Rizzolatti and Sinigaglia, op. cit., p. 124.
17. *The Empathic Brain: How the Discovery of Mirror Neurons Changes Our Understanding of Human Nature* by Christian Keysers, published in paperback by Social Brain Press, 2011, pp. 41–44.
18. Ibid., pp. 95–98.
19. "Both of us disgusted in *my* insula: the common neural basis of seeing and feeling disgust" by B. Wicker, C. Keysers, J. Plailly, J.P. Royet, V. Gallese and G. Rizzolatti, published in *Neuron*, vol. 40, pp. 655–664, 2003.
20. Keysers, op. cit., p. 98.
21. "A common anterior insula representation of disgust observation, experience and imagination shows divergent functional connectivity pathways" by Mbemba Jabbi, Jojanneke Bastiaansen and Christian Keysers, published in *PLoS ONE*, vol. 3, issue 8, e2939, August 2008.
22. Keysers, op. cit., p. 108.
23. Iacoboni, op. cit., p. 256.
24. "Brain regions with mirror properties: a meta-analysis of 125 human fMRI studies" by Pascal Molenberghs, Ross Cunnington and Jason B. Mattingley, published in *Neuroscience and Biobehavioral Reviews*, vol. 36, 2012, p. 348.
25. "Single-neuron responses in humans during execution and observation of actions" by Roy Mukamel, Arne D. Ekstrom, Jonas Kaplan, Marco Iacoboni, and Itzhak Fried, published in *Current Biology*, vol. 20, 27 April 2010, p. 6.
26. "Evolution of mirror neuron mechanism in primates" by Giacomo Rizzolatti and Leonardo Fogassi, published in *Evolution of Nervous Systems*, 2nd edition, vol. 3, 2017, pp. 373–386.
27. Ibid., pp. 380–382.
28. "Automatic imitation" by Cecilia Heyes, published in *Psychological Bulletin*, vol. 137, no. 3, 2011, pp. 463–483, p. 478.

29. Ibid., p. 479.
30. *The Myth of Mirror Neurons: The Real Neuroscience of Communication and Cognition* by Gregory Hickok, published by W.W. Norton & Company, Inc., 2014, pp. 184–206.
31. Ibid., p. 206.
32. Rizzolatti and Fogassi, op. cit., pp. 381–382.
33. Hickok, op. cit., pp. 49–52.
34. "Grasping the intentions of others with one's own mirror neuron system" by Marco Iacoboni, Istvan Molnar-Szakacs, Vittorio Gallese, Giovanni Buccino, John C. Mazziotta and Giacomo Rizzolatti, published in *PLoS Biology*, vol. 3, e79, 2005, pp. 4–5.
35. *Brainstorming: Views and Interviews on the Mind* by Shaun Gallagher, published by Imprint Academic, 2008, pp. 124–134.
36. "The functional role of the parieto-frontal mirror circuit: interpretations and misinterpretations" by Giacomo Rizzolatti and Corrado Sinigaglia, published in *Nature*, vol. 11, April 2010, p. 271.
37. "Mirror neurons and motor intentionality" by Giacomo Rizzolatti and Corrado Sinigaglia, published in *Functional Neurology*, vol. 22, 2007, pp. 205–210.
38. "Before and below theory of mind: embodied simulation and the neural correlates of social cognition" by Vittorio Gallese, published in *Philosophical Transactions of the Royal Society B Biological Sciences*, vol. 362, 2007, pp. 659–669.
39. "Mirror in action" by Corrado Sinigaglia, published in *Journal of Consciousness Studies*, vol. 16, 2009, pp. 309–334.
40. Hickok, op. cit., pp. 41–76.
41. Hickok, op. cit., p. 43.
42. "Neural circuits underlying imitation learning of hand actions: an event-related fMRI study" by Giovanni Buccino, Stefan Vogt, Afra Ritzl, Gereon R. Fink, Karl Zilles, Hans-Joachim Freund and Giacomo Rizzolatti, published in *Neuron*, vol. 42, 22 April 2004, p. 323.
43. "Cortical mechanisms underlying the organization of goal-directed actions and mirror neuron-based action understanding" by Giacomo Rizzolatti, Luigi Cattaneo, Maddalena Fabbri-Destro and Stefano Rozzi, published in *Physiological Reviews*, vol. 94 (2), April 2014, p. 683.

Chapter 17. The Mirrored Homunculus

1. *Soul Dust: The Magic of Consciousness* by Nicholas Humphrey, published by Quercus, 2011, p. 5.
2. Blackmore, op. cit., p. 51.
3. "'Reality' is constructed by your brain. Here's what that means, and why it matters' by Brian Resnick, published on https://www.vox.com/science-and-health/20978285/optical-illusion-science-humility-reality-polarization, 22 June 2020.
4. "Illusionism as a theory of consciousness" by Keith Frankish, published in *Journal of Consciousness Studies*, vol. 23 (11–12), 2016, pp. 11–39, p. 30.
5. Frankish, op. cit.
6. Frankish, op. cit, p. 22.
7. Humphrey, op. cit., p. 40.
8. Humphrey, op. cit., p. 61.
9. *The Language of Thought: A New Philosophical Direction* by Susan Schneider, published by MIT Press, 2011, pp. 45–46.
10. Koch, op. cit., p. 135.
11. Schore, op. cit., p. 82.
12. *Winnicott* by Adam Phillips, published in paperback by Penguin Books, 2007, pp. 128–129.
13. Phillips, op. cit., p. 3.
14. Phillips, op. cit., p. 4.
15. "Residential management as treatment for difficult children" by Donald Winnicott and Clare Britton, 1947, cited in *Deprivation and Delinquency* by Donald Winnicott, published by Tavistock, 1984, p. 58.
16. Ramachandran, op. cit., p. 260.
17. "Philosophical conceptions of the self: implications for cognitive science" by Shaun Gallagher, published in *Trends in Cognitive Sciences*, vol. 4, no. 1, January 2000, p. 14.
18. Ibid., p. 16.
19. Phillips, op. cit., p. 121.
20. Iacoboni, op. cit., p. 134.
21. Phillips, op. cit., p. 4.
22. Phillips, op. cit., pp. 4–5.
23. *The Better Angels of Our Nature: A History of Violence and Humanity* by Steven Pinker, published by the Penguin Group, 2011, p. 590.

Chapter 18. Testing the Mirrored Homunculus Theory

1. *The Consciousness Instinct: Unravelling the Mystery of How the Brain Makes the Mind* by Michael S. Gazzaniga, published by Farrar, Strauss and Giroux, 2018, p. 103.
2. Gopnik, op. cit.
3. Ibid., p. 145.
4. *Animal Languages: The Secret Conversations of the Living World* by Eva Meijer, published by John Murray (Publishers), 2019, pp. 73–74.
5. "Fish passes mirror test for first time" by Yvaine Ye, published in *New Scientist*, 8 September 2018.
6. Gopnik, op. cit., pp. 59, 148.
7. Ibid., pp. 54–61.
8. Ibid., p. 147.
9. Ibid., pp. 147–150.
10. Ibid., pp. 150–152.
11. "Oxytocin modulates social value representations in the amygdala" by Yunzhe Liu, Shiyi Li, Wanjun Lin, Wenxin Li, Xinyuan Yan, Xuena Wang, Xinyue Pan, Robb B. Rutledge and Yina Ma, published in *Nature Neuroscience*, vol. 22, April 2019, pp. 633–641.
12. "Serotonin-mediated inhibition of ventral hippocampus is required for sustained goal-directed behaviour" by Keitaro Yoshida, Michael R. Drew, Masaru Mimura and Kenji F. Tanaka, published in *Nature Neuroscience*, vol. 22, May 2019, pp. 770–777.
13. "Inside the mind's eye" by Daniel Cossins in *New Scientist*, 8 June 2019.
14. "Attentional and self-regulatory difficulties of Romanian orphans ten years after being adopted to Canada: a longitudinal study" by Karyn Audet, submitted as a thesis at the Simon Fraser University, September 2003.
15. Ibid., p. iii.
16. https://www.humanbrainproject.eu/en/about/overview/.

Chapter 20. What Human-Purposed AI Looks Like

1. Bostrom, op. cit., p. 209.
2. *The Value Learning Problem* by Nate Soares, Machine Intelligence Research Institute, 2015, p. 7.

3. *The History of Philosophy* by A.C. Grayling, published by Viking, 2019, pp. 264–265.
4. https://en.wikipedia.org/wiki/Trolley_problem.
5. "How moral are you? No, really?" by Sylvia Terbeck, published in *New Scientist*, 31 October 2020.
6. Grayling, op. cit., p. 281.
7. *The Righteous Mind: Why Good People are Divided by Politics and Religion* by Jonathan Haidt, published by Allen Lane, 2012, p. 26.
8. Haidt, op. cit., pp. 191–192.
9. *No Man is an Island* by John Donne, 1572–1631.
10. "Why do we care" by Patricia Churchland, published in *New Scientist*, 28 September 2019.
11. Dignum, op. cit., p. 41.
12. "A theory of cultural value orientations: explication and applications" by S. Schwartz, published in *Comparative Sociology*, vol. 5, issue 2, 2006, pp. 137–182.
13. *Culture's Consequences: Comparing Values, Behaviours, Institutions and Organizations* by Geert Hofstede, published by Sage Publications, 2001.
14. "Artificial intelligence, values, and alignment" by Iason Gabriel, published in *Minds and Machines*, vol. 30, 1 October 2020, pp. 411–437.
15. "'Second thoughts': what makes North Korean defectors want to go back?" by Justin McCurry, 16 January 2022, published on https://www.theguardian.com/world/2022/jan/16/second-thoughts-what-makes-north-korean-defectors-want-to-go-back.
16. *The New Breed: How to Think about Robots* by Kate Darling, published by Allen Lane, 2021, p. 163.
17. "Building machines that learn and think for themselves: commentary on Lake, Ullman, Tenenbaum, and Gershman, Behavioral and Brain Sciences 2017" by M. Botvinick, D.G.T. Barrett, P. Battaglia, N. de Freitas, D. Kumaran, J.Z. Leibo, T. Lillicrap, J. Modayil, S. Mohamed, N.C. Rabinowitz, D.J. Rezende, A. Santoro, T. Schaul, C. Summerfield, G. Wayne, T. Weber, D. Wierstra, S. Legg, and D. Hassabis, published by *DeepMind* on https://arxiv.org/ftp/arxiv/papers/1711/1711.08378.pdf, abstract.

18. Quoted in *Brainstorming: Views and Interviews on the Mind* by Shaun Gallagher, published by Imprint Academic, 2008, p. 168.
19. "Super-intelligent machines" by Bill Hibbard, published in *Computer Graphics*, 2001, vol. 35(1), pp. 11–13.
20. Darling, op. cit., p. 158.
21. "Is there an elegant universal theory of prediction?" by Shane Legg, published in *Algorithmic Learning Theory, ALT 2006, Lecture Notes in Computer Science*, vol. 4264, edited by J.L. Balcázar, P.M. Long and F. Stephan, published by Springer, 19 October 2006, p. 10.

Chapter 21. Building the Sentient Robot

1. *Machines Like Us* by Ian McEwan, published by Jonathan Cape, 2019, pp. 110–118.
2. Taken from an interview with Cynthia Breazeal, published in Ford, op. cit., p. 451.
3. "Robot emotion: a functional perspective" by Cynthia Breazeal and Rodney Brooks, published as chapter 10 of *Who Needs Emotions: The Brain Meets the Robot*, edited by Jean-Marc Fellous and Michael A. Arbib, published by Oxford University Press, 2005, p. 279.
4. Ibid., p. 302.
5. Ibid., p. 306.
6. "Emotions: from brain to robot" by Michael A. Arbib and Jean-Marc Fellous, published in *Trends in Cognitive Sciences*, vol. 8, no. 12, December 2004, p. 554.
7. *The Demon in the Machine: How Hidden Webs of Information are Solving the Mystery of Life* by Paul Davies, published by Allen Lane, 2019, p. 68.
8. "Ethical artificial intelligence" by Bill Hibbard, published on arxiv.org/abs/1411.1373, 27 January 2016, p. 89.
9. Ibid., pp. 78–92.
10. Ibid., p. 90.
11. *In Our Own Image: Will Artificial Intelligence Save or Destroy Us?* by George Zarkadakis, published by Ebury Publishing, 2015, p. 286.
12. "Intel to release neuromorphic-computing system" by Sara Castellanos, published on https://www.wsj.com/articles/intel-

to-release-neuromorphic-computing-system-11584540000, 18 March 2020.

13. "Artificial intelligence accelerated by light" by Huaqiang Wu and Qionghai Dai, published in *Nature*, vol. 589, 6 January 2021, pp. 25–26.
14. *What is Quantum Computing* by CB Insights, 2020.
15. "A quantum trick with photons gives AI a speed boost" by Leah Crane in *New Scientist*, 20 March 2021.
16. "The architectural basis of affective states and processes" by Aaron Sloman, Ron Chrisley and Matthias Scheutz, published in Fellous and Arbib, op. cit., chapter 8, p. 239.
17. Domingos, op. cit., pp. 223–227.
18. *The Deep History of Ourselves: The Four-Billion-Year Story of How We Got Conscious Brains* by Joseph LeDoux, published by Viking, 2019, p. 229.
19. Domingos, op. cit., pp. 227–233.
20. *The AI Does Not Hate You* by Tom Chivers, published by Weidenfeld & Nicolson, 2019, p. 66.
21. "Improving the accuracy of medical diagnosis with causal machine learning" by Jonathan G. Richens, Ciarán M. Lee and Saurabh Johri, published in *Nature Communications*, 11 August 2020.
22. Domingos, op. cit., p. 239.
23. Domingos, op. cit., p. 25.
24. Domingos, op. cit., p. 250.
25. Cantwell Smith, op. cit., p. 146.
26. Cantwell Smith, op. cit., p. 132.
27. "A plan to advance AI by exploring the minds of children" by Will Knight, published in *MIT Technology Review*, 12 September, 2018 on www.technologyreview.com/s/612002/a-plan-to-advance-ai-by-exploring-the-minds-of-children.
28. "Should humanity build a global AI Nanny to delay the singularity until it's better understood" by Ben Goertzel, contained in *The Singularity: Could Artificial Intelligence Really Out-Think Us (and Would We Want It to)?*, edited by Uziel Awret, published by Imprint Academic, 2016, p. 163.

29. "Leakproofing the singularity: artificial intelligence confinement problem" by Roman V. Yampolskiy, published in Awret, op. cit., p. 263.
30. *Artificial You: AI and the Future of Your Mind* by Susan Schneider, published by Princeton University Press, 2019, p. 51.

Epilogue

1. "The singularity: a philosophical analysis" by David J. Chalmers, 2010, published in Awret, op. cit., p. 45.

Index